Peter D Riley

Cambridge
checkpoint

NEW EDITION

checkpoint
Science
3

Workbook

HODDER EDUCATION
AN HACHETTE UK COMPANY

Hachette UK's policy is to use papers that are natural, renewable and recyclable products and made from wood grown in sustainable forests. The logging and manufacturing processes are expected to conform to the environmental regulations of the country of origin.

Orders: please contact Bookpoint Ltd, 130 Milton Park, Abingdon, Oxon OX14 4SB. Telephone: (44) 01235 827827. Fax: (44) 01235 400454. Lines are open 9.00–5.00, Monday to Saturday, with a 24-hour message answering service. Visit our website at www.hoddereducation.com.

© Peter D Riley 2013
First published in 2013 by
Hodder Education, an Hachette UK Company,
Carmelite House, 50 Victoria Embankment,
London EC4Y 0DZ

Impression number   9
Year   2018

All rights reserved. Apart from any use permitted under UK copyright law, no part of this publication may be reproduced or transmitted in any form or by any means, electronic or mechanical, including photocopying and recording, or held within any information storage and retrieval system, without permission in writing from the publisher or under licence from the Copyright Licensing Agency Limited. Further details of such licences (for reprographic reproduction) may be obtained from the Copyright Licensing Agency Limited, Saffron House, 6–10 Kirby Street, London EC1N 8TS.

Cover photo © Gusto images/Science Photo Library
Typeset in Palatino Light 10.5/12.5 by Integra Software Services Pvt. Ltd., Pondicherry, India
Printed in Great Britain by CPI Group (UK) Ltd, Croydon CR0 4YY

A catalogue record for this title is available from the British Library

ISBN 978 1444 183 504

Contents

	Introduction	iv
	BIOLOGY	**1**
Chapter 1	Photosynthesis	1
Chapter 2	Reproduction in flowering plants	9
Chapter 3	Adapting to a habitat	18
Chapter 4	Ecosystems	27
Chapter 5	Human influences on the environment	32
Chapter 6	Classification and variation	40
	CHEMISTRY	**46**
Chapter 7	The structure of the atom	46
Chapter 8	The periodic table	49
Chapter 9	Endothermic and exothermic reactions	53
Chapter 10	Patterns of reactivity	59
Chapter 11	Preparing common salts	63
Chapter 12	Rates of reaction	66
	PHYSICS	**71**
Chapter 13	Density	71
Chapter 14	Pressure	76
Chapter 15	Turning on a pivot	81
Chapter 16	Electrostatics	84
Chapter 17	Electricity	87
Chapter 18	Heat energy transfers	93
Chapter 19	World energy needs	98

Introduction

What does an anther make? What is a top carnivore? What do the parts of an atom look like? Which element explodes on contact with water? What does a catalyst do? How do you calculate density? What happens to make a spark? Where does geothermal energy come from?

These are some of the questions you will be asked in this book to see if you are becoming a 'scientific citizen'. They will also help you prepare for your Cambridge Checkpoint Science Tests.

The chapters here follow the sequence of the chapters in the Checkpoint Science Student's Book 3. There are headings within each chapter here to help you relate them to chapters and sections in the student's book, so you can integrate your work.

Most of the questions aim to test your knowledge and understanding of science but some questions have this icon . These questions aim to test your science enquiry skills. There is one of these questions in almost every chapter and in some chapters there may be two.

So now it is time to start. Read each question carefully, think about it, then write down the answer in the space provided or as instructed. Often you will have to write down some facts or explanations, sometimes you will need to tick a box and occasionally you will have to link details by lines or construct and interpret graphs.

At the end of each chapter there is a space for your teacher to write comments to help you with your progress.

> # BIOLOGY

1 Photosynthesis

Starch in leaves

1. What is the natural form of starch in a leaf? Tick (✔) the correct box.

   Colourless crystals ☐

   Brown crystals ☐

   Colourless grains ☒

   Black grains ☐

2. Akram is testing a leaf for starch.

   a) What items should he select from this list of apparatus? Circle your answers.

   - (beaker)
   - Liebig condenser
   - Bunsen burner
   - test tube
   - (gauze)
   - conical flask
   - microscope
   - heat-proof mat
   - (white tile)
   - (clamp and stand)
   - filter funnel
   - crucible
   - tripod

   b) He will need water to perform **three** tasks. What are they? List them in order.

   1. _to boil to leaf and make iodine easier to enter cells._

   2. _boil ethanol to take chlorophyll out of leaf_
      water
   3. _to leaf in to make it softer._

   Akram is using ethanol in his test.

CHAPTER 1

c) (i) What does he need it for? _remove chlorophyll from leaf_

(ii) What safety precautions must he take when using ethanol?

Make sure to turn bunsen burner off and wear gloves and safety googles

Akram completed the first part of the test and the leaf is ready for testing for starch.

d) (i) What chemical reagent does Akram need to see if starch is present?

iodine

(ii) What colour is the reagent in the bottle? _clear (orange-brownish)_

(iii) What colour does the reagent go if starch is present in the leaf?

blue-black

Jamila points out to Akram that he has used a leaf from a plant that has been in a cupboard for 2 days.

e) (i) What result does Jamila predict for Akram's test?

that not much starch will be present

(ii) Explain your answer. _the leaf hasn't had sunlight to produce glucose and store it as starch._

Constructing the equation for photosynthesis

3 Han is planning an investigation to show that plants need water to stay healthy.

a) What does he need for his investigation? _two plants witch have been in the same conditions, same age, same type and water_

b) What should he do? _He should water one of the plants with very few water and the other with a considerable amount for about 1 week – 1 month_

c) What result do you predict? _I predict that the plant that was watered more will have produced more glucose._

4 Budi has set up this plant as half of an investigation.

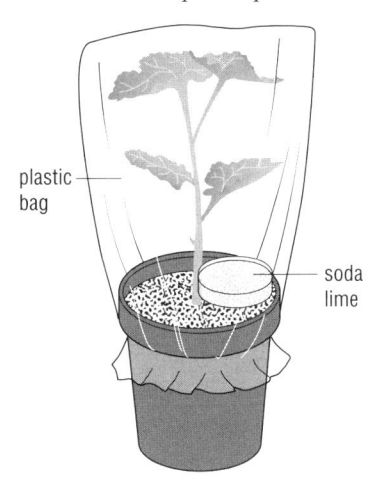

a) If the investigation is to work, what must Budi have done to the plant?

_____

b) What is the purpose of the soda lime? _To absorb all CO_2 in the bag._

c) What must Budi do to set up the other half of his investigation?

Set up a plant with the same conditions, but with sodium hydrogencarbonate solution

d) Predict what Budi may find when he makes his starch tests at the end of the investigation. _The plant with soda lime will have no starch and the one with sodium hydrogencarbonate will have starch._

CHAPTER 1

5 A plant has been growing in a sunny garden and Asoka tests one of its leaves for starch.

a) What will be the result? _The plant will have starch_

Joko tidies up the garden and puts a plant pot over the plant by mistake. Two days later Asoka tests another of its leaves for starch.

b) (i) What will be the result? _the plant will not have starch_

(ii) Explain your answer. _it will not be able to do photosynthesis, since it has no sunlight._

6 a) What is a variegated leaf? _a leaf which doesn't have chlorophyll everywhere_

b) Why is a variegated leaf used in photosynthesis investigations? _to test the effect of chlorophyll on photosynthesis_

Asoka tests the leaf of a variegated plant growing in the sunny garden.

c) (i) What does she find when she tests the leaf for starch?
Tick (✔) the correct box.

The leaf is yellow and blue. ☐

The leaf is brown and blue-black. ✓

The leaf is red and brown. ☐

The leaf is red and blue-black. ☐

(ii) Why does the test show **two** colours? _because green parts do photosynthesis and white don't, since there's no chlorophyll._

PHOTOSYNTHESIS

7 Tahirah set up the experiment shown in the diagram.

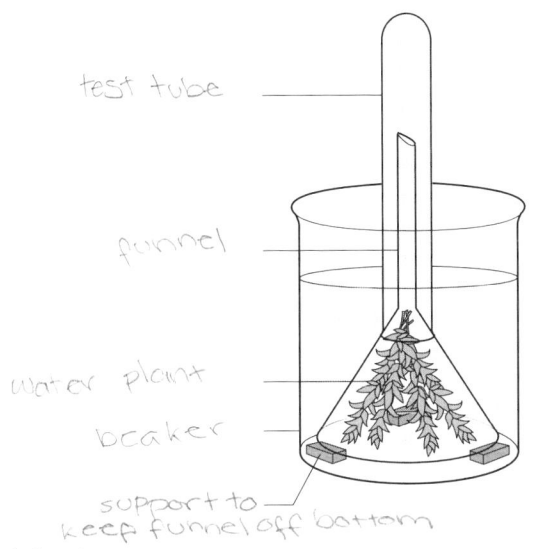

- test tube
- funnel
- water plant
- beaker
- support to keep funnel off bottom

a) Label the diagram.

She puts it on a sunny windowsill and leaves it for a week.

b) Describe the appearance of the experiment after 1 week. _there will be gas at the top of the test tube._

c) (i) What could Tahirah do to take her experiment further?

Set up the same experiment twice, and put one in the dark and one in sunlight

(ii) Predict her result. _the one in the dark will have no gasses at the top of the test tube,_

8   Ali and Lily set up some data loggers around plants to measure the amount of light, carbon dioxide and oxygen. The graphs show the changes in the amounts over a day, starting at midnight.

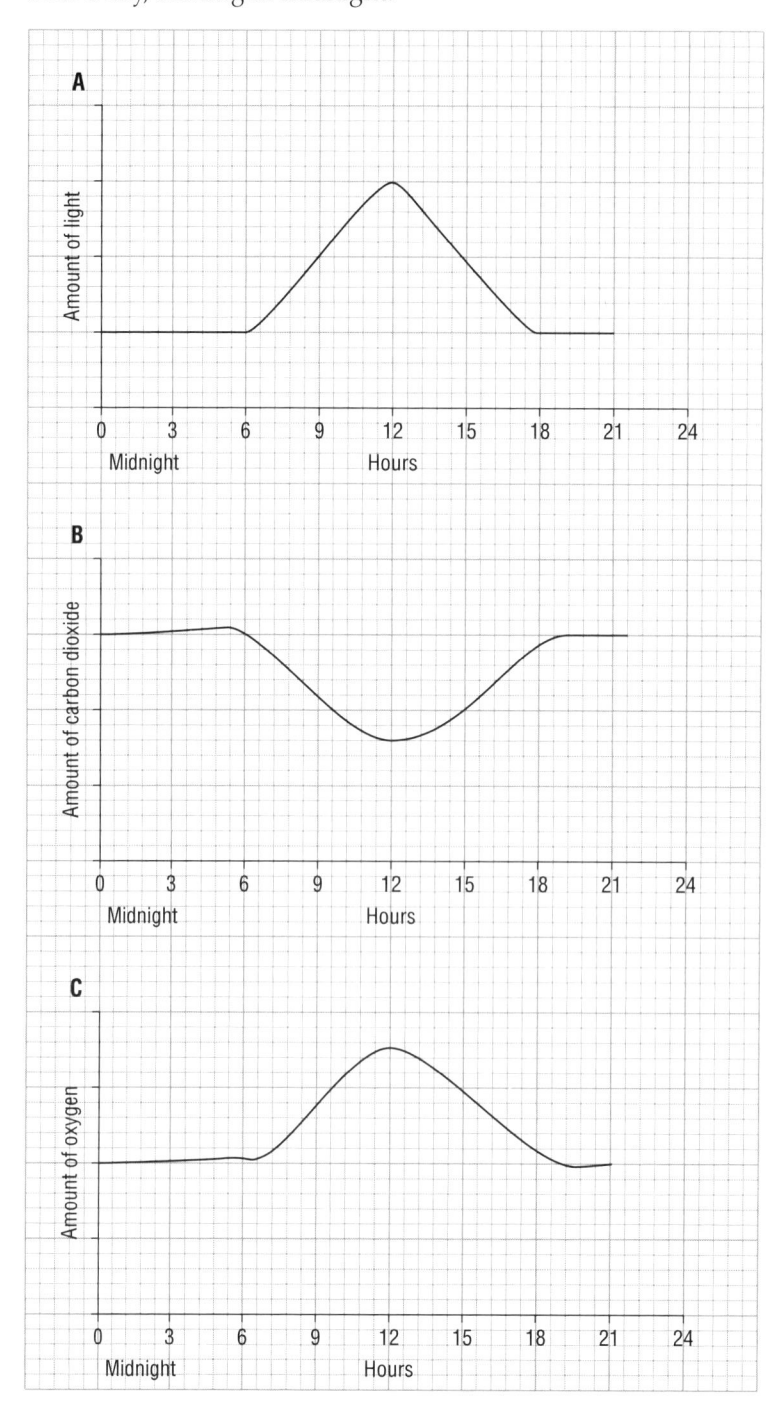

A

Amount of light

0 3 6 9 12 15 18 21 24
Midnight   Hours

B

Amount of carbon dioxide

0 3 6 9 12 15 18 21 24
Midnight   Hours

C

Amount of oxygen

0 3 6 9 12 15 18 21 24
Midnight   Hours

a) Describe the pattern shown in graph A. _light appears during hours 6 to 18 since its the only time the plant can do photosynthesis_

b) Describe the pattern shown in graph B. _amount of CO_2 decreases during hours 6 to 18 since the plant is absorbing it to do photosynthesis because there's sunlight_

c) Compare the pattern of graph C with graph A. _amount of O_2 increases during hour 6 to 18 because there is sunlight for photosynthesis_

d) Suggest how the patterns of the graphs may be linked together.

Graph A shows when there's light for photosynthesis, therefore the plant absorbs CO_2 during that time and produces O_2.

9 a) Draw lines to match the formulae of the chemicals with their common names.

Formula	Common names
CO_2	water
$C_6H_{12}O_6$	oxygen
H_2O	carbon dioxide
O_2	carbohydrate

● CHAPTER 1

b) Use the common names to write a word equation for photosynthesis and include **two** other essential factors above and below the arrow.

$$\text{water} + \text{carbon dioxide} \xrightarrow[\text{chlorophyll}]{\text{sunlight}} \text{glucose} + \text{oxygen}$$

Mineral salts

10 a) Which of these elements do plants obtain from mineral salts? Circle your answers.

- oxygen
- (nitrogen)
- hydrogen
- (phosphorus)
- carbon
- (potassium)

b) Which mineral is needed for:

(i) healthy root growth phosphorus (P)

(ii) healthy flower growth potassium (K)

(iii) healthy leaf growth? nitrogen (N)

c) What happens to the minerals in a plant when it dies?

They are released and after used to create new living organisms

Teacher comments

2 Reproduction in flowering plants

The parts of a flower

1   The diagram shows an insect-pollinated flower.

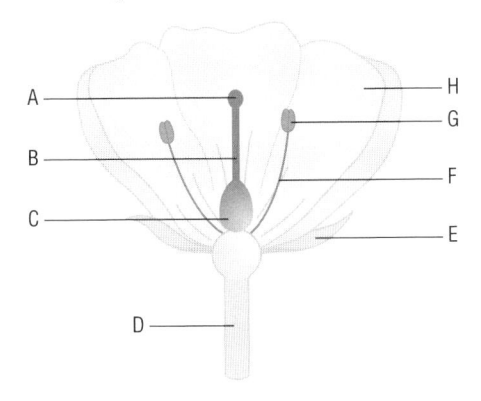

Here are the names of its parts.

- sepal
- petal
- ovary
- stigma
- stamen
- style
- stalk
- anther

a) Write the name of each part in the correct place in the table.

Part	Name
A	stigma
B	style
C	ovary
D	stalk
E	sepal
F	filament/stamen
G	anther
H	petal

● CHAPTER 2

b) Which part protects the flower when it is in bud? _sepal_

c) Which part forms part of the corolla? _petals_

d) (i) Which parts form the male part of the flower? _filament + anther_

(ii) What is the name of the male part of the flower? _stamen_

e) (i) Which parts form the female part of the flower? _ovary, style, stigma_

(ii) What is the name of this part? _carpel_

2 a) What is nectar? _a sweet liquid for insects to feed on_

b) What is the name for the parts that produce it? _nectaries_

c) Where are these parts found? _under the ovary, on top of the stalk_

3 Which statements are true about wind-pollinated flowers?
Tick (✔) the two correct boxes.

They produce large quantities of nectar.	☐
They produce large quantities of pollen.	✔
Their stigmas hang outside their flowers.	✔
They are larger than insect-pollinated flowers.	☐

Pollen grains and pollination

4 a) What does a pollen grain contain? _the male sex cells_

b) Where is pollen produced? _in the anthers_

c) Some pollen grains are covered in spikes.
   (i) What kind of flowers produce them? __insect-pollinated__

   (ii) What is the advantage of them having spikes? __They can stick to their body easily__

d) Some pollen grains are small and light in weight.

   (i) What kind of flowers produce them? __wind-pollinated__

   (ii) What is the advantage of them being light in weight? __the wind can carry them easily.__

5 a) What happens in the process of pollination? __the pollen travels from anthers to stigmas__

   b) How can a flower avoid self-pollination? Tick (✔) the two correct boxes.

   Release pollen when stigmas can receive them. ☐

   Release pollen before stigmas are ready to receive them. ☒

   Release pollen after stigmas are ready to receive them. ☒

   Have male and female parts in one flower. ☐

6 There is a beehive in the school grounds. The students want to investigate bee behaviour. Dembe predicts that the bees are attracted to all colours of flowers equally. The class sets out to test this prediction.

They make four model flowers out of card and paint each one a different colour – red, blue, white and yellow. In the centre of each flower they place a pot to hold a solution, which is a substitute for nectar.

a) How must the flowers be constructed to make the test fair?

they should be the same size and have the same amount of nectar

b) (i) Which solid and liquid could be used to make a solution like nectar?

___

(ii) Which physical process takes place in the making of the solution?

___

Usen suggests using a different concentration of nectar solution for each flower.

c) Why do the others say it is not a good plan? _because the bees might like one better and not get attracted by the colour, but by the nectar solutions_

The students take the model flowers out towards the hive.

d) How should they place the flowers to make the test fair? Explain your ideas.

We should place them the same distance away from the hive because bees might see some flowers first, or feel safer closer to the hive.

The students take turns to record how many times bees visit flowers in 2 minute periods. They record their results as tally marks in the table.

Name	Blue flower visits in 2 minutes	Red flower visits in 2 minutes	White flower visits in 2 minutes	Yellow flower visits in 2 minutes
Dembe	/////	/	///// ///	///// ////
Usen	///	//	///// ///	///// ///
Ochi	////	/	///// /	///// ///// /
Totals	12	4	22	28

e) Complete the totals in the table.

f) Select the best type of graph – a bar graph or line graph – and construct a graph of the data.

REPRODUCTION IN FLOWERING PLANTS

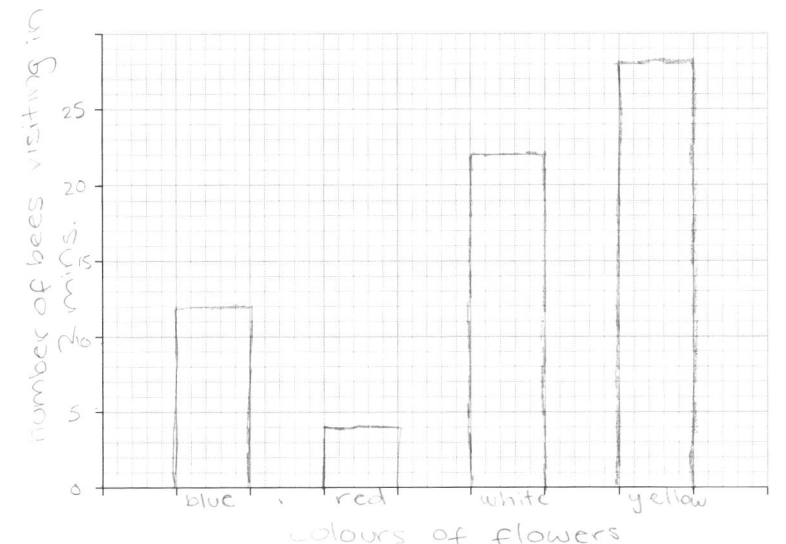

[Bar chart: y-axis "number of bees visiting in 5 mins" (0 to 25+); x-axis "colours of flowers" with bars: blue ~12, red ~4, white ~22, yellow ~28]

g) Does the data support Dembe's prediction? Explain your answer.

No, because there is a big difference in # of bees

h) How could the investigation be modified to make the results more reliable?

Using real flowers

Fertilisation

7 The following sentences describe the stages in fertilisation but they are in the wrong order. Arrange them correctly by writing the letter of each statement in the order in which it occurs.

A Male gamete enters the ovule.

B Male gamete moves into pollen tube.

C Pollen grain opens.

D Male and female gametes fuse together.

E Pollen tube forms.

F Male gamete leaves tip of pollen tube.

G Zygote is produced.

E C B F A D G

● CHAPTER 2

8 After fertilisation what happens to:

a) the ovule _it turns into the seed_

b) the petals and stamens _fall away_

c) the ovary? _it turns into the fruit_

Dispersing the fruits and seeds

9 a) Why are seeds dispersed? _So that they do not compete with the parent plant for nutrients, space, water and light_

b) Why do plants produce a large number of fruits and seeds?
Tick (✔) the correct box.

To increase the chance that some will find a place to grow	✔
To provide food for animals	☐
To help recycle minerals	☐
To use up energy stored in the leaves	☐

10 Joko and Nidhi have a collection of seeds. They have six of each type shown in the picture.

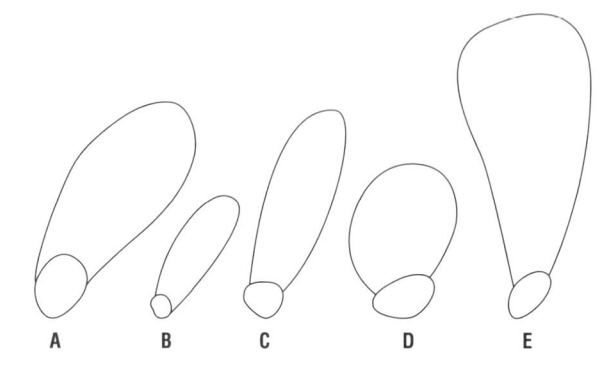

A   B   C   D   E

They believe there is a relationship between the wing size and the distance that the seed travels.

a) (i) What do you think they predict? _That the bigger the wing, the further the seed will travel._

REPRODUCTION IN FLOWERING PLANTS

(ii) Give an explanation for the prediction. _The larger the wing, the more air it can catch._

Here is a list of equipment that is available in the laboratory.

- microscope
- (hairdryer)
- scissors
- (ruler)
- Bunsen burner
- clamp and stand
- tripod
- gauze
- slide
- forceps
- mounted needle
- (tape measure)
- centrifuge
- beaker

b) Circle the items in this list that they should use for the investigation.

c) Which statement describes the appearance of the seeds?
Tick (✔) the correct box.

The seeds and wings are all the same size. ☐

The seeds are all the same size but the wings are different sizes. ☐

The seeds and wings are all different sizes. ☒

The larger the seed, the larger the wing. ☐

d) Write a plan for Joko and Nidhi's investigation.

1. Measure all the wing sizes. 2. Set up tape measure to measure distance travelled. 3. Place seed at the start of tape measure. 4. Turn hair dryer on and measure distance. Repeat with all seeds.

● CHAPTER 2   GERMINATION: seed begins to grow

11 A scientist has a set of chambers in her laboratory, set to different temperatures. She puts a tray containing 50 seeds of the same plant in damp soil in each chamber and records the number germinated after 5 days.

Here are her results.

Temperature/°C	Number of seeds germinated
0	0
5	2
10	5
15	10
20	20
25	30

a) Make a line graph from the data in this table.

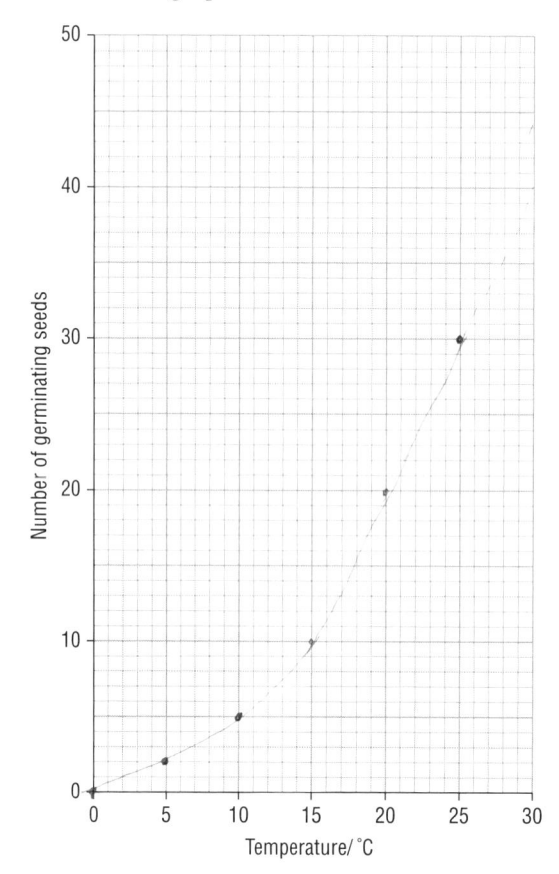

b) In this investigation, name:

   (i) the dependent variable _number of seeds germinated_

   (ii) the independent variable. _temperature (°C)_

c) Use the graph to predict the number germinating at 30 °C. _40 seeds_

d) (i) If a tray of seeds was set up at 100 °C, predict how many would germinate. _None_

   (ii) Explain your answer. _100 °C is not a natural temperature and all water would be turned to water vapour, not being able to be used._

Teacher comments

3 Adapting to a habitat

1   a) The soil water in a habitat collects deep below the surface. How may a species of plant become adapted to reach it?

   It may have long *and deep* roots to spread out and reach the water.

   b) A species of plant lives in the habitat of large leaf-eating herbivores, such as deer. How may its leaves become adapted to help the species survive?

   They may develop to have poison in them or spikes on their surface.

   c) A species of bird feeds on worms on muddy river banks. The worms adapt by burrowing deeper. How might the bird species become adapted so that they can keep feeding on the worms? They may develop a longer beak

Adaptations to the seasons

2   The table shows the temperature and rainfall records for a 12-month period in one habitat.

Weather feature	Jan	Feb	Mar	Apr	May	Jun	Jul	Aug	Sep	Oct	Nov	Dec
Temperature/°C	37	35	32	28	24	22	22	25	29	32	34	36
Rainfall/mm	45	33	28	12	10	9	6	7	8	20	31	40

ADAPTING TO A HABITAT

a) Make a line graph of the data about the temperature.

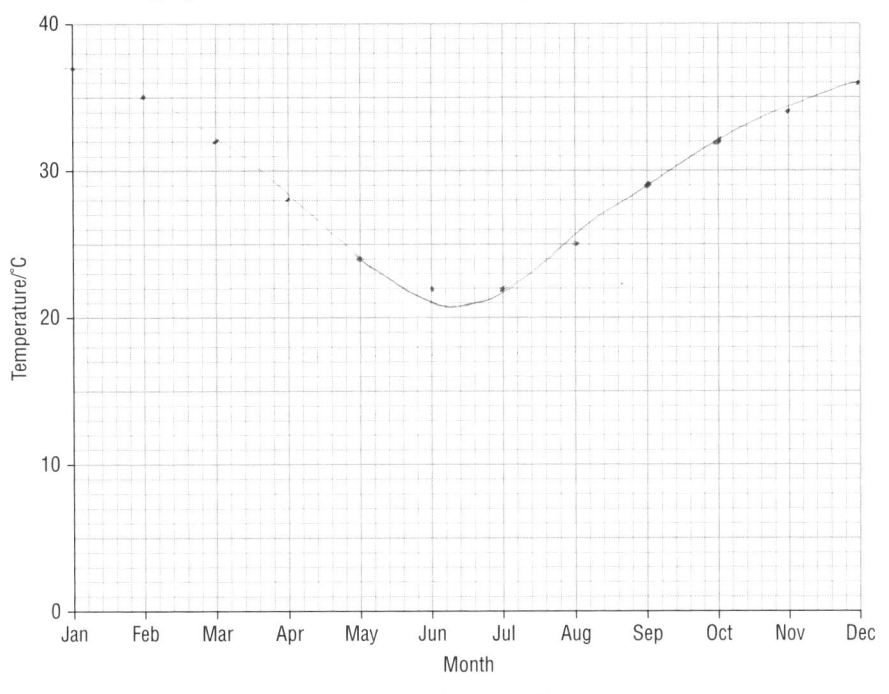

b) Make a bar graph of the data about the rainfall.

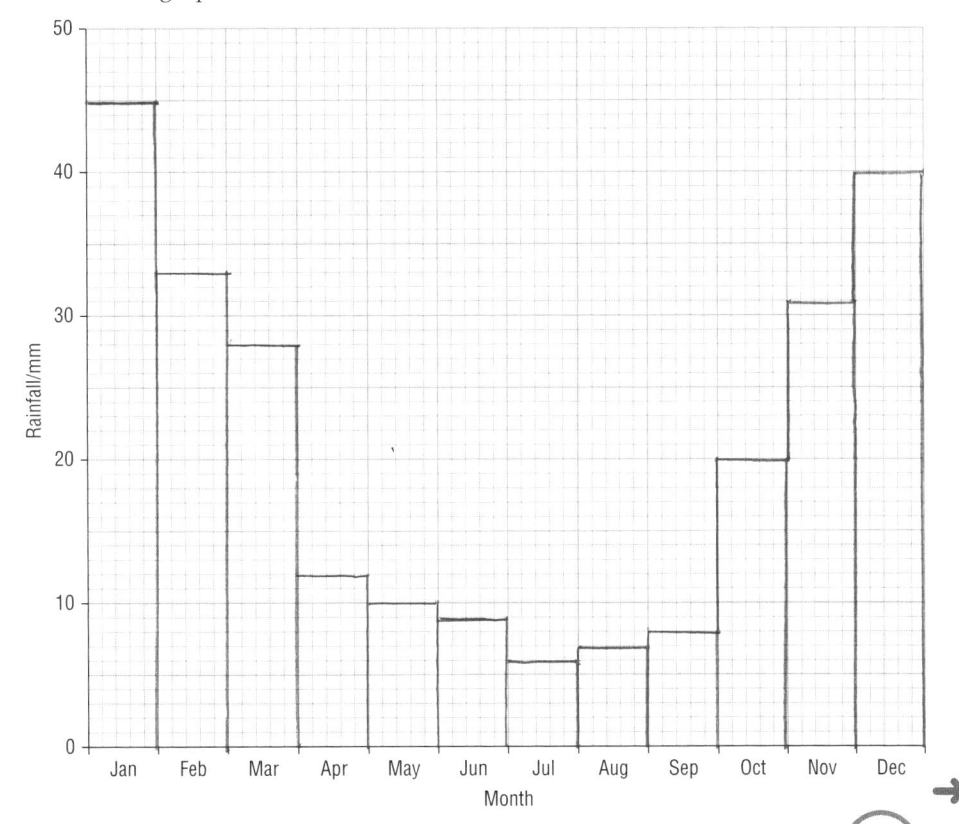

● CHAPTER 3

c) (i) In which hemisphere is the habitat? __lower hemisphere__

(ii) Explain your answer. _____

d) (i) How many seasons does it have? __2__

(ii) Explain your answer. __there are 6 months where there is very few rainfall (12-7 mm), and 6 months where there is more rainfall (20-45 mm)__

e) Look at the millimetre scale on a ruler and consider the rainfall in this habitat. Which kind of habitat do you think it is? Circle your answer.
- rainforest
- desert
- polar region
- (woodland)

f) Which types of organisms would be adapted to this habitat? Circle your answers.
- tree dwellers, like monkeys
- camel-like animals
- animals with white fur and feathers
- tree-nesting birds
- mosses and lichens
- very tall trees
- cactus-like plants
- ferns and other woodland plants

3 Here are 12 statements about the activities of organisms in a European wood over a year. Place them in the correct season by writing the letter of each statement in the appropriate box in the table.

A Hibernating animals wake up.

B Leaves of deciduous trees lose chlorophyll.

C Deciduous trees are without leaves.

D Trees produce nuts and berries.

E Hibernating animals build up stores of fat.

F The woodland is in shade and has fewer flowers.

20

G Some insectivorous animals hibernate.

H Plants with bulbs grow leaves and flowers.

I Caterpillars feed on leaves.

J Many insects survive as eggs and pupae.

K Deciduous trees grow leaves.

L Birds lay second or third clutches of eggs.

Spring	Summer	Autumn	Winter
A D	I	B E	C

F, G, H, J, K, L

4 Why do animals migrate? _because they have not adapted to survive in their previous habitat during the following season._

Adaptations to a habitat

5 Plants can grow in mangrove swamps and in swamps at the edge of lakes. The two habitats have different features.

   a) Tick (✔) the habitat to which the feature belongs.

Feature	Mangrove swamp	Lake
salt water	✔	
fresh water		✔
daily rise and fall of water levels	✔	
seasonal rise and fall of water levels		✔
risk of drying out in drought		✔
no risk of drying out in drought	✔	
no movement of mud		✔
regular movement of mud	✔	

   b) Describe how a plant must be adapted to survive in a swamp by a lake.

● CHAPTER 3

6 Firoza has been looking at books about birds in their habitats. She has noticed that most birds have three forward-facing toes and one backward-facing toe and that in woodland birds, the back toe is longer than in birds which live on grasslands.

Firoza thinks that the long back toe helps the birds to perch and decides to set up an experiment to test her idea. She realises that she cannot use real birds and thinks about making simple models of them. She knows that wire is sometimes used to make models and decides to use it to make her birds. The picture shows a 'bird' about to be placed on its perch.

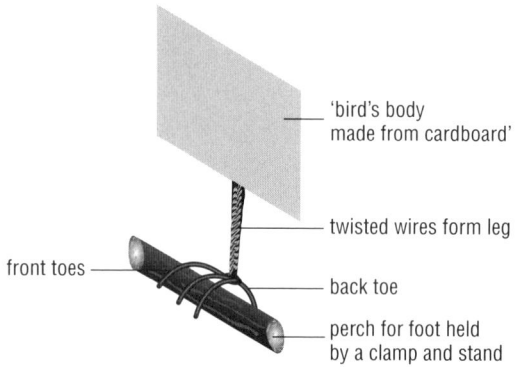

- 'bird's body made from cardboard'
- twisted wires form leg
- front toes
- back toe
- perch for foot held by a clamp and stand

a) What evidence did Firoza consider when setting up her scientific enquiry?

_____

b) What creative thought occurred to Firoza as she tried to devise an experiment?

_____

c) Firoza also used a hairdryer in this investigation. How do you think she used it? _____

_____

Firoza makes three model 'birds' to test.

d) Which parts of the models do you think she kept the same?

_____

_____

e) Which part of the model do you think she changed?

f) Which bird do you think held on to the perch most easily?

g) Which bird fell off its perch most easily?

Extreme adaptations

7. A mouse and a lizard are hiding in the same dark hole when a pit viper arrives at the entrance.

   a) Name **three** sense organs that the pit viper can use to detect prey.

   b) Which sense organ does it use to detect prey in the dark hole?

   c) (i) Which of the two animals do you think the snake detects more easily?

   (ii) Explain your answer.

   d) Why can the snake judge distance to make an accurate strike?

   e) What kills the prey when the snake strikes?

8 A class of students investigated plants growing under a tree. They set up a line transect with six places, called stations, on it where they collected data.

a) What is a line transect? _____

_____

The figure below shows the location of the stations.

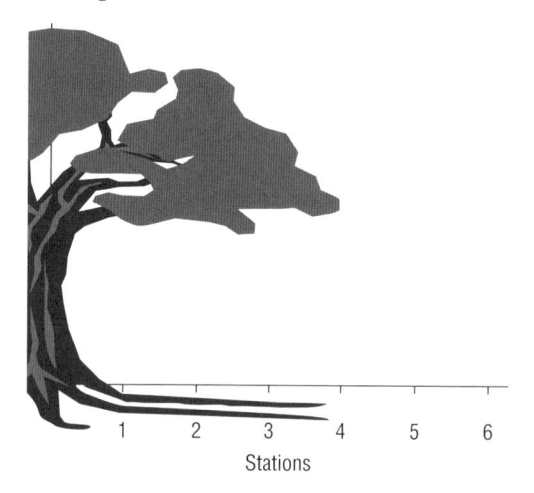

They used a data logger to record the amount of light at each of the stations and produced a graph as shown below.

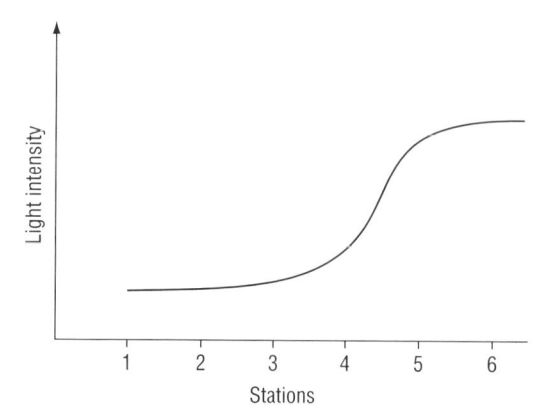

They used a quadrat at each station to record the species and number of plants.

b) What is a quadrat? _____

_____

The table shows their results.

Station	Plant species A	Plant species B
1	0	1
2	1	4
3	2	4
4	8	3
5	15	1
6	16	0

c) Use the data to make bar graphs for each species in the space below. The axes have been prepared for you.

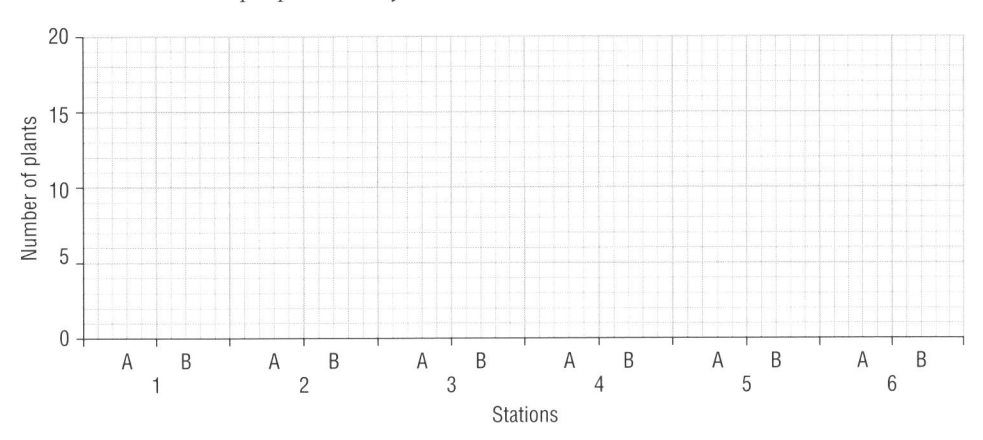

d) Which statements describe the relationship between the first figure and the presentation of the data in your graphs? Tick (✔) the two correct boxes.

As the light increases the number of species A decreases. ☐

As the light increases the number of species B decreases. ☐

As the light increases the number of species A increases. ☐

As the light increases the number of species B increases. ☐

e) (i) Why do you think plant A was absent from station 1?

_____

(ii) Why do you think plant B was absent from station 6?

_____

Ravi says that the distribution of the plants is due entirely to the way they are adapted to light.

f) (i) Is he correct? _____

(ii) Explain your answer. _____

_____

_____

Teacher comments

4 Ecosystems

The growth of ecology and A vocabulary of ecology

1 a) Which continent is in the Nearctic realm? _____

   b) Which continent is in the Neotropical realm? _____

2 a) Rearrange the following words and add a few more to write a sentence to explain what biome means.
   - Earth
   - plants
   - large region
   - same community
   - generally covered
   - all parts
   - having same weather
   - animals

   _____

   _____

   b) What are the **two** largest biomes in Africa?

   _____

   c) What is the major biome of the Australasian realm?

   _____

3 Which realm has the smallest land mass? _____

4 What does **biodiversity** mean? _____

   _____

5 Name the **three** major life forms in a community.

   _____

CHAPTER 4

6 Which are the abiotic factors in this list? Circle your answers.
- light
- disease-causing organisms
- wind frequency
- soil type
- temperature
- large numbers of predators
- humidity
- presence of rocks
- small numbers of prey

Food chains and Food webs

7 A mongoose is a large, weasel-like mammal that can be found in Africa, Asia, southern Europe, some Caribbean islands and the Hawaiian islands.

   a) In which biogeographical realms is it found?

   b) Here is one food chain of the mongoose.

   grass → grasshopper → lizard → mongoose

   Write the name of the organism from the food chain in the appropriate box in the table. Some terms apply to more than one organism.

Ecological term	Organism
producer	
primary consumer	
herbivore	
secondary consumer	
tertiary consumer	
carnivore	
top carnivore	
prey	
predator	

8 Two components of the food chains and food webs in the sea are the phytoplankton and zooplankton. Phytoplankton are plant-like microorganisms called algae and zooplankton are the larvae of crustaceans, molluscs and worms. Krill is a small, shrimp-like animal.

   a) Here are some animals. Circle those that are crustaceans and underline those that are molluscs.
   - spider
   - centipede
   - snail
   - wasp
   - shrimp
   - millipede
   - slug
   - lobster
   - octopus
   - crab
   - jellyfish

   Here is a simple food web of the organisms in the sea.

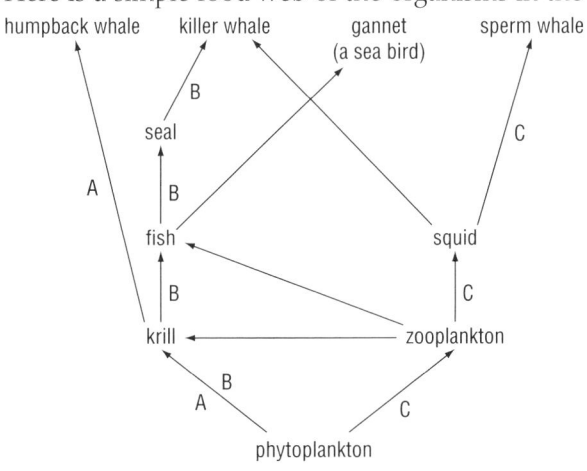

   b) Identify the feeding status of the whale in food chains A–A, B–B–B–B and C–C–C by placing a tick (✔) in an appropriate box in the table.

Whale	Secondary consumer	Tertiary consumer	Quaternary consumer
humpback			
killer			
sperm			

   c) Identify the food chain with **six** organisms in it and write it out here.

   _____

   _____

   d) Why might a killer whale eat more squid if the numbers of fish decrease?

   _____

   _____

   e) What is the energy source for this food web? _____

● CHAPTER 4

Ecological pyramids and Decomposers

9 **a)** In the space provided, draw a triangle to represent a pyramid and label it to represent four links in the food chain.

b) Two chimpanzees are using sticks to collect and eat 100 termites, which are feeding on a rotten log. In the space provided, construct and label a pyramid of numbers using boxes to indicate the number of individuals at each tier.

Ecosystems

10 Match each group of organisms with their activities by putting a tick (✔) in the appropriate boxes in the table.

Organisms	Take in carbon dioxide	Produce carbon dioxide	Take in oxygen	Produce oxygen	Release minerals
producers					
primary consumers					
secondary consumers					
decomposers					

How populations change

11 Draw lines to match each pair of birth rate and death rate with its effect on the population.

Birth rate	Death rate	Effect on population
10 per 1000	20 per 1000	population does not change
20 per 1000	20 per 1000	population increases
15 per 1000	10 per 1000	population decreases

12 a) How can zoos increase the birth rate of endangered animals?

_____

b) How can zoos decrease the death rate of endangered species?

_____

Teacher comments

5 Human influences on the environment

Humans in the environment

1 The first humans used the natural materials that were around them. Draw lines to match the materials with their uses.

Material	Use
shells	clothes
stone	fuel
wood	building material
flint	containers
skins	knives

2 The Industrial Revolution increased demand for two things to make larger numbers of products.

   a) Which **two** things were in greater demand? _____

   b) Name **two** ways in which the environment was affected. _____

   c) What caused the increased demand for products in the Industrial Revolution?

HUMAN INFLUENCES ON THE ENVIRONMENT

The Earth's changing atmosphere

3 Carbon dioxide was present in the earliest atmosphere.

   a) Where did it come from? _____

   The earliest plants used the carbon dioxide in a food-making process.

   b) What is this process? _____

   c) What is the waste product of this process? _____

   Ultraviolet rays from the Sun react with some of the waste product.

   d) What substance is produced? _____
   e) How is this substance beneficial to life on Earth?

   _____

   f) How has this substance been damaged by human activity?

   _____

   _____

Air pollution

4 Which of the following substances are produced by the burning of oil and coal? Circle your answers.
   - sodium chloride
   - carbon dioxide
   - calcium carbonate
   - oxides of nitrogen
   - carbon monoxide
   - copper sulfate
   - sulfur dioxide
   - helium

5 The following are stages in the warming of the atmosphere due to 'greenhouse' gases. They are in the wrong order. Arrange them correctly by writing the letter of each statement in the order in which it occurs.

   A Some heat is released from the Earth's surface.

   B The heat passes through space from the Sun.

   C Some heat is released from the atmosphere into space.

   D The Sun releases heat.

   E Some heat is absorbed by the atmosphere and warms it.

   F Some heat is absorbed by the Earth.

   G Heat passes through the atmosphere.

   _____

6 a) What are the substances produced by burning fuels that form acid rain?

   _____

   b) Name **two** acids that they produce. _____

   _____

   c) When acid rain enters a pond, how does the pH change?
      Tick (✔) the correct box.

      The pH goes down.   ☐

      The pH goes down and up.   ☐

      The pH goes up.   ☐

      The pH goes up and down.   ☐

   d) Here is a food chain in the pond.
      algae → small invertebrates → fish
      The small invertebrates are killed by the acid rain.

      (i) What do you predict will happen to the population of algae? Explain your

          answer. _____

      _____

HUMAN INFLUENCES ON THE ENVIRONMENT

(ii) What do you predict will happen to the population of fish? Explain your answer. _____

_____

e) (i) Which metal is leached from the soil by acid rain? _____

(ii) Which aquatic animals does an excess of this metal affect?

_____

(iii) What happens to these animals? _____

7 Gimbya is studying the amount of soot brought by the wind into a habitat. When the wind blows in a certain direction, she leaves out a white board and collects specks of soot on it. The diagram shows the boards that collected soot when the wind blew from different parts of the compass.

a) Count up the specks on each board and complete the table.

Board	Number of specks of soot
wind from north	
wind from south	
wind from east	
wind from west	

b) Construct a bar graph from the data in the table.

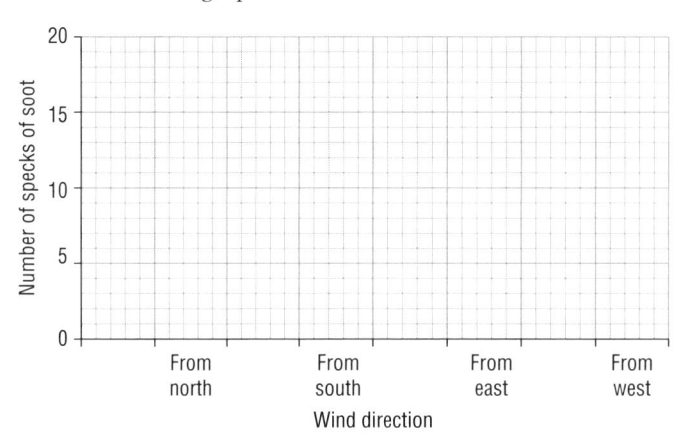

c) (i) In which direction is the source of soot pollution? _____

(ii) Explain your answer. _____

Akilah says that the test might not be fair.

d) (i) Do you agree? _____

(ii) Explain your answer. _____

Water pollution

8 Name **three** industries that have caused water pollution in rivers.

_____

_____

9 a) What is an algal bloom? Tick (✔) the correct box.

   A small flowering of algae plants. ☐

   A small increase in the algae population. ☐

   A great increase in the algae population. ☐

   A great flowering of algae plants. ☐

   b) What causes an algal bloom? Tick (✔) the correct box.

   Acid rain ☐

   Fertiliser ☐

   PCBs ☐

   Mercury ☐

   c) After an algal bloom, what takes in large quantities of oxygen? Tick (✔) the correct box.

   Fish ☐

   Algae ☐

   Bacteria ☐

   Freshwater shrimps ☐

   d) What happens as a result of an algal bloom? _____

   _____

10 State **two** ways in which the 'load on top' cleaning method for oil tankers helps the environment.

   1 _____

   2 _____

● CHAPTER 5

Indicators of pollution

11 A factory releases harmful gases into the air and they blow through a forest close by. Scientists investigate the effect of the air pollution by examining lichens growing on the trees. They make a transect through the forest, recording the lichen types as they move towards the factory. Station 1 is furthest from the factory.

Patterns	Station 1	Station 2	Station 3	Station 4	Station 5	Station 6	Station 7
A	crusty	crusty	leafy	leafy	bushy	bushy	none
B	bushy	bushy	leafy	leafy	crusty	crusty	none
C	bushy	leafy	bushy	crusty	leafy	bushy	none

a) Which one of the patterns shown in the table do you think they identified?

b) The factory stopped releasing harmful gases. At which stations would you predict a change in the lichen population? _____

c) How do you think the lichen population of the forest will change after several years of clean air blowing through it? _____

Intensive farming

12 Identify the facts about the two types of fertiliser by ticking (✔) the appropriate boxes in the table.

Fact	Inorganic fertiliser	Organic fertiliser
contains manufactured chemical compounds		
needs activity of decomposers		
contains animal and human wastes		
provides minerals instantly		
does not need activity of decomposers		
provides minerals slowly		
builds up soil structure		
can be added while crop is growing		
constant use destroys soil structure		
cannot be added to crop while it is growing		

HUMAN INFLUENCES ON THE ENVIRONMENT

13 Why are weeds harmful to crops? _____

_____

14 There are two kinds of insecticides – broad-spectrum insecticides and narrow-spectrum insecticides. Which one makes less impact on the environment?

Explain your answer. _____

_____

15 What does the term **biological pest control** mean? _____

_____

16 A farmer has set up a glasshouse to grow crops. She has installed some burners to produce extra carbon dioxide, blinds on the inside of the glass and heaters.

a) Why does she need them? _____

_____

b) Why may she have to alter the setting of each piece of equipment when she changes her crops? _____

_____

c) What else must she supply to the plants if they are to grow well?

_____

Teacher comments

6 Classification and variation

The main groups of living things

1. Draw lines to match each kingdom with its features and an example.

Kingdom	Features	Example
Protoctista	have spores and hyphae	bacteria
Fungi	one cell body without nucleus	amoeba
Monera	one cell body with nucleus	yeast

2. Damisi made some notes about the plants she saw on a field trip. To which group does each belong?

   a) Green slime. _____

   b) Roots, stems, fronds and spores. _____

   c) Woody stem and cones. _____

3. Haroon visits an aquatic habitat and makes some notes about the plants and animals he sees there.

   a) To which group does each organism belong?

   (i) Soft body, tentacles at one end around the mouth _____

   (ii) Does not have roots, stems or leaves _____

   (iii) Spiny skin and globe shaped _____

   (iv) Feathers and beak _____

CLASSIFICATION AND VARIATION

b) What kind of aquatic habitat do you think Haroon visited?

c) Which organism spent the least amount of time in the water?

Keys

4   The fins of a fish are shown in the diagram.

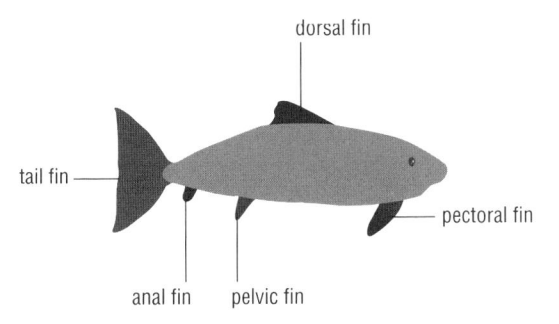

Here are four fish found in the North Sea on the coast of Europe.

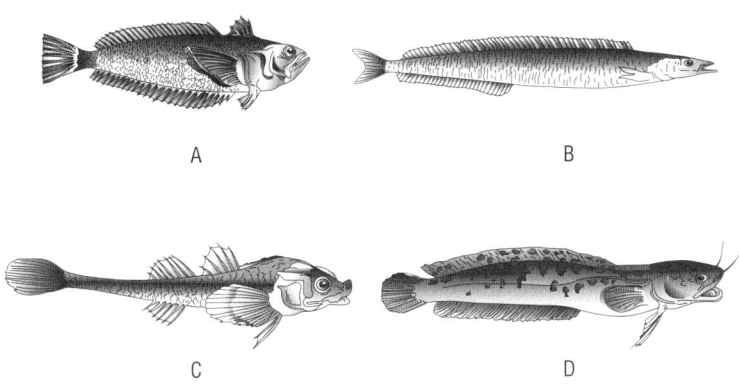

A

B

C

D

Here is a key to identify the fish.

1	Fish with dorsal fin in two parts	go to 2
	Fish with one long dorsal fin	go to 3
2	Small anal fin	Armed bullhead
	Long anal fin	Lesser weever
3	Large pectoral and pelvic fins	Three-bearded rockling
	Small pectoral and pelvic fins	Lesser sand-eel

a) Use the key to identify each fish.

Fish A is _____

Fish B is _____

Fish C is _____

Fish D is _____

b) One fish is poisonous – it has a tail fin with a straight edge. What is its name?

_____

c) One fish is the food of many sea birds – it has a forked tail. What is its name?

_____

Inherited characteristics

5   A characteristic is an observable feature that is always found in a particular type of organism. What is an inherited characteristic? _____

_____

Cells and reproduction

6   In which part of the cell are chromosomes found? Tick (✔) the correct box.

Cytoplasm   ☐

Vacuole   ☐

Chloroplast   ☐

Nucleus   ☐

7 Which **two** of the following statements are correct? Tick (✔) the two correct boxes.

There are two genes for each body feature in a cell. ☐

There are two chromosomes for each body feature in a cell. ☐

There is one gene on each chromosome. ☐

There are large numbers of genes on a chromosome. ☐

8 a) What are the gametes or sex cells of animals called?

_____

b) Where are the male gametes of flowering plants found?

_____

c) Where are the female gametes of flowering plants found?

_____

The male gamete of an animal has 12 chromosomes.

d) How many chromosomes are in the gametes of a female animal of the same species? _____

e) These male and female gametes join together. What is this process called?

_____

f) What is formed as a result of this process and how many chromosomes does it have? _____

9 a) During the formation of chromosomes in gametes, what happens to the chromosomes? _____

b) Why does this chromosome activity affect the appearance of the offspring of the parents? _____

● CHAPTER 6

Selection

10 Scientists have been asked to selectively breed a wheat plant that can produce a crop of seeds for harvesting in a cold, windy place with low rainfall. Here are some features of plants that could be used in the breeding programme.

  A Long stem
  B Long roots
  C Short roots
  D Short stem
  E Photosynthesises only above 10 °C
  F Photosynthesises only above 5 °C
  G Seeds leave stalk early when fully formed
  H Seeds cling to stalk when fully formed.

Select the features you think they need and give a reason for your choice.

_____

_____

_____

_____

11 a) Which statement describes the result of natural selection? Tick (✔) the correct box.

  The strongest of two species will survive in a habitat. ☐

  The best suited of two species will survive in a habitat. ☐

  The more aggressive of two species will survive in a habitat. ☐

  The more numerous of two species will survive in a habitat. ☐

b) Which process helps a species change over time to survive in a habitat? Tick (✔) the correct box.

Predation ☐

Adaptation ☐

Conservation ☐

Respiration ☐

12 What does the evolution of a species mean? Rearrange the following parts of a sentence to make the definition and write it below.
- over time
- natural selection
- may change
- of
- through the process
- a species

_____

_____

Teacher comments

CHEMISTRY

7 The structure of the atom

1. What does the word **atom** mean? Tick (✔) the correct box.

   Tiny particle ☐   Indivisible particle ✔

   Invisible particle ☐   Made of particles ☐

2. Which statement describes what happens in a chemical reaction? Tick (✔) the correct box.

   Matter is created. ☐

   Matter is destroyed. ☐

   Matter is created then destroyed. ☐

   Matter is neither created nor destroyed. ✔

The structure of atoms

3. a) Draw lines to match the subatomic particle to its charge.

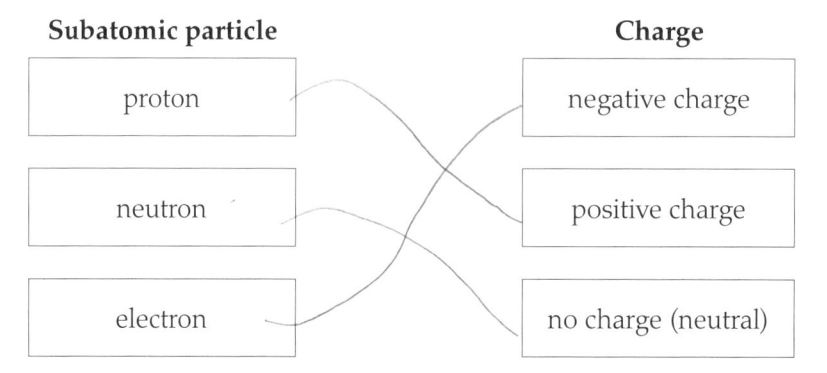

Subatomic particle	Charge
proton	negative charge
neutron	positive charge
electron	no charge (neutral)

proton → positive charge; neutron → no charge (neutral); electron → negative charge

THE STRUCTURE OF THE ATOM

b) What is the charge of the atom? Circle your answer.
- positive
- negative
- (neutral)

4 The diagram shows the structure of a beryllium atom.
   a) Label the parts.

   A electron
   B shells
   C nucleus

   b) (i) Which is the part that is positively charged? __nucleus__

   (ii) Explain your answer. It contains protons, which are positively charged. (Also neutrons which have a neutral charge.)

5 Which of these statements about an atom of an element is true?
Tick (✔) the two correct boxes.

It always has the same number of protons.	✔
It always has the same number of neutrons.	☐
The number of protons can vary.	☐
The number of neutrons can vary.	☐

6 How fast does an electron travel? Tick (✔) the correct box.

About the speed of sound	☐
About the speed of light	✔
Supersonic speed	☐
Speed of a tortoise	☐

● CHAPTER 7

The atomic structure of 20 elements

7 Draw lines to match the element's name with its symbol and atomic structure.

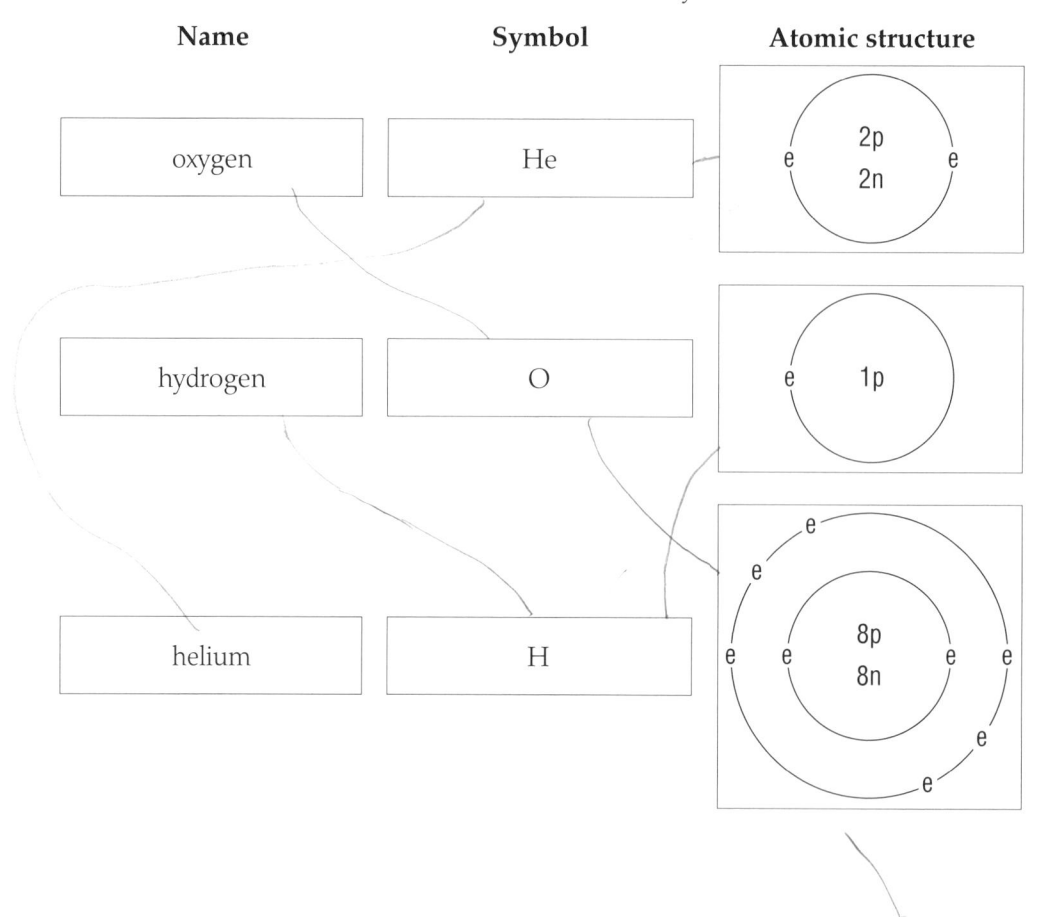

Name	Symbol	Atomic structure
oxygen	He	(e, 2p 2n, e)
hydrogen	O	(e, 1p)
helium	H	(8p 8n, with electrons e around inner and outer shells)

Teacher comments

8 The periodic table

The periodic table

1 Which statement describes the atomic number? Tick (✔) the correct box.

The number of electrons in an atom ☐

The number of protons in an atom ☑

The number of neutrons in an atom ☐

The number of protons and neutrons in an atom ☐

2 a) Identify the elements missing from the following groups. Give their names, symbols and atomic numbers in the table below.

Group	Name	Symbol	Atomic number
1	Potassium	K	19
2	Beryllium	Be	4
4	Silicon	Si	14
5	Nitrogen	N	7
7	Flourine	F	9
8	Argon	Ar	18

● CHAPTER 8

Groups of the periodic table

3 Look at the periodic table in the previous question to help you with this one. Write the number of the group that is known as the:

   a) halogens ___7___

   b) alkaline earth metals ___2___

   c) alkali metals. ___1___

4 a) Identify the group to which each of the following elements belongs from the data about their melting points and boiling points.

Element	Melting point/°C	Boiling point/°C	Group	Group name
A	649	1097	2	Alkali earth metals
B	−100.9	−34	7	Halogens
C	63.5	760	1	Alkali metals

   b) What is the name of the type of properties featured in the table?

   _____

5 Identify the group to which each of the following elements belongs from the data about their chemical properties.

Element	Chemical property	Group	Group name
A	dissolves in water to form an acid	7	Halogens
B	dissolves slightly in water to form an alkaline solution	2	Alkali earth metals
C	dissolves readily in water to form an alkaline solution	1	Alkali metals

THE PERIODIC TABLE

6 a) Identify each of these elements and their groups from the information about them.

Element	Information	Name	Group
A	pale yellow-green poisonous gas		
B	forms bones and teeth	Ca	1 – Alkali metals
C	combines with oxygen, silicon, and aluminium to form an emerald		
D	used to make batteries	Li	1 – Alkali metals
E	forms the orange glow in some street lamps		
F	red-brown liquid with poisonous fumes		

b) To which groups do the following elements belong?

A: 8 protons in its atoms ___group 6___

B: 14 electrons in its atoms ___group 4___

C: 13 protons in its atoms ___group 3___

7 a) Draw lines to match these elements to their uses.

Element	Use
helium	airport landing lights
krypton	wire filament light bulbs
xenon	meteorological balloons
argon	photographers' flash guns

b) What is the name of the group to which all these elements belong?

___8 – noble gasses___

c) What chemical property do they all share?

they are very unreactive

8 a) What is the appearance of hydrogen? Circle your answer.
- pale green
- pale yellow
- dark blue
- violet
- (colourless)
- pale pink

b) What compound does hydrogen form with nitrogen?

ammonia

c) In crude oil, what compound does hydrogen form with carbon?

hydrocarbons

The periods of the periodic table

9 a) Which statement best describes a period in the periodic table? Tick (✔) the correct box.

A vertical line of elements ☐

A diagonal line of elements ☐

A horizontal line of elements ✔

A circle of elements ☐

b) (i) Which period has the most non-metals? _2_

(ii) Name them. _carbon, nitrogen, oxygen, flourine_

c) In the part of the periodic table that you have studied, which period has the largest number of metals? _6_

Teacher comments

9 Endothermic and exothermic reactions

Endothermic reactions

1 a) Heat is taken in when ice melts but is melting an endothermic *chemical* reaction? _no, it is a endomerthic physical process_

   b) Explain your answer. _there is no new chemical compound being created, only a change in state_

2 a) Rearrange these substances to make a word equation for the reaction that occurs when you put a sherbet sweet in your mouth.
   - water
   - citric acid
   - carbon dioxide
   - sodium citrate
   - sodium hydrogen carbonate

   citric acid + sodium hydrogen-carbonate → sodium citrate + carbon dioxide + water

   b) What substance makes the sherbet fizz? _carbon dioxide_

3 a) What is the chemical compound from which limestone is made?

   calcium carbonate

   b) (i) How many elements form this compound? _3_

      (ii) What are their symbols? _$CaCO_3$_

      (iii) When limestone is heated in a kiln, which substances are produced?

      calcium oxide + carbon dioxide

   c) What is burnt in a limestone kiln to provide the heat? _gas_

CHAPTER 9

4 Which statement suggests that photosynthesis is an endothermic reaction? Tick (✔) the correct box.

Raising the temperature increases the rate of photosynthesis. ☒

Lowering the temperature increases the rate of photosynthesis. ☐

Any change of temperature increases the rate of photosynthesis. ☐

No change in temperature increases the rate of photosynthesis. ☐

Exothermic reactions

5 Which of these words describes a chemical reaction in which a substance reacts quickly with oxygen and heat is given out? Tick (✔) the correct box.

Combination ☐

Condensation ☐

Combustion ☒

Concentration ☐

6 This apparatus is used to investigate the products of a burning candle.

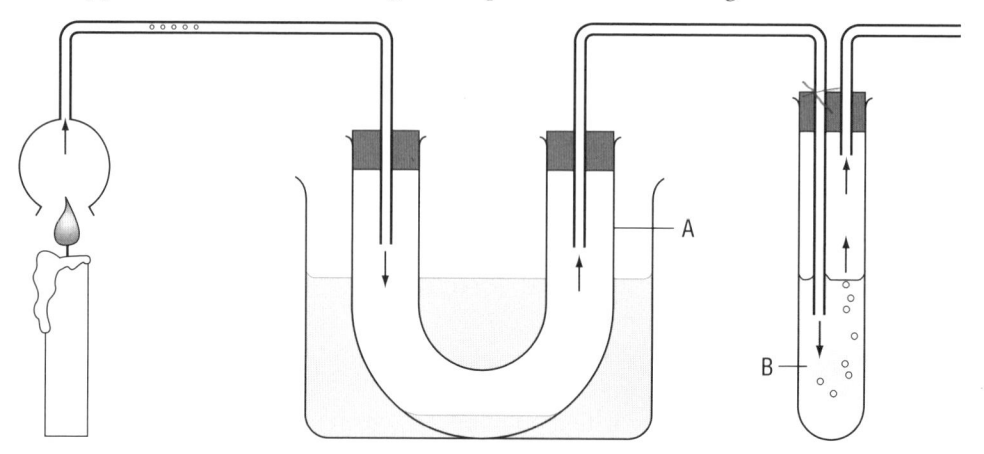

a) (i) What is used to make the current of air flow through the apparatus?

a suction pump

(ii) Mark with an X where this device is attached to the apparatus.

ENDOTHERMIC AND EXOTHERMIC REACTIONS

b) (i) What process occurs in A? __condensation__

    (ii) What substance is produced? __water__

    (iii) What material can be used to identify it? __cobalt chloride__

    (iv) What colour change occurs in this material to positively identify the substance? __it changes from blue to pink__

c) (i) What reagent is placed in B? __lime water__

    (ii) What substance is this reagent used to identify? __carbon dioxide__

    (iii) What change occurs if this substance is present? __the lime water turns cloudy / milky__

The candle is made from a hydrocarbon.

d) Construct the word equation to describe the burning of a candle using the word 'hydrocarbon'.

__oxygen + hydrocarbon → carbon dioxide + water__

7 Rafiq is finding out how much energy is in a quantity of candle wax. He sets up the apparatus shown in the diagram. He records the mass of the candle on a balance and records the temperature of the 500 cm³ of water in the beaker. He lights the candle, stirs the water and records the water temperature until it has risen 10 °C. He puts out the burning candle and records its mass again.

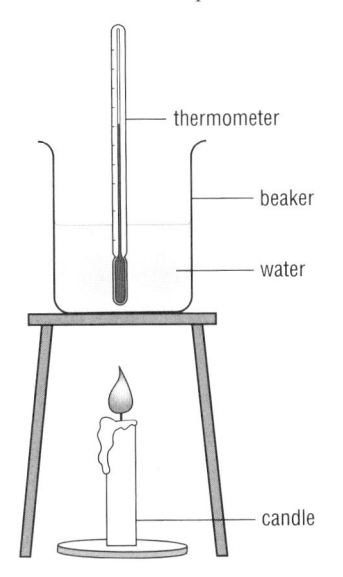

Here are his results.

Mass of candle before burning	10 g
Mass of candle after burning	5 g
Loss of mass	5 g

Temperature before heating	22 °C
Temperature after heating	32 °C
Rise in temperature	10 °C

a) Complete the two tables.

The energy in 1 g of candle is found by using the formula below:

$$\frac{2.1 \times \text{rise in temperature}}{\text{loss of mass}} = \text{kJ/g}$$

b) Use the formula to find the energy in the candle wax that was burnt away by Rafiq.

$$\frac{2.1 \times 10}{5} = \frac{21}{5} = 4.2$$

4.2 kJ/g

8 Jaya is comparing the heat produced by two fuels. She uses each one in turn to heat the same volume of water for 10 minutes and records the temperature every 2 minutes. The table shows her data.

Time/minutes	Fuel A temperature/°C	Fuel B temperature/°C
0	20	20
2	22	24
4	24	28
6	26	32
8	28	36
10	30	40

a) Draw and label a line graph for each fuel.

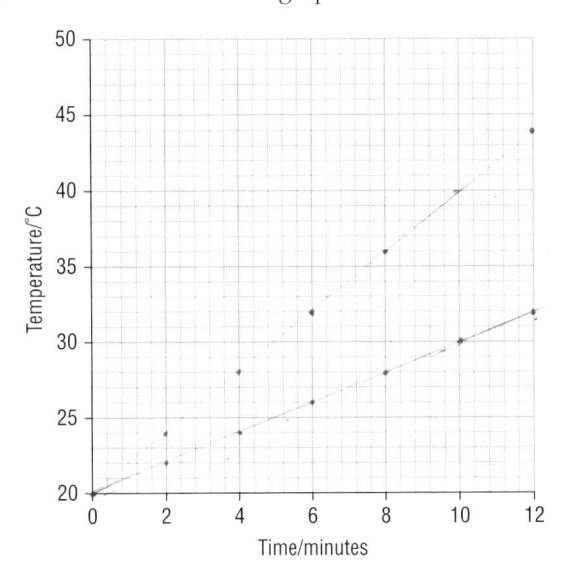

b) Use the graph to predict what the temperature of the water would have been after 12 minutes. Mark your prediction with a dot.

c) Using this predicted data, predict the temperature difference between the two beakers of water after 12 minutes. ___12°C___

9 glucose + oxygen → carbon dioxide + water

  a) What is the process described by this word equation? ___respiration___

  b) Where does it take place? ___In all living things___

  c) The first reactant is made in an endothermic reaction. Name the reaction.

  ___photosynthesis___

  d) Name **two** sources of the second reactant used in the process named in part **a**.

  ___air, released in photosynthesis___

  e) What is the energy released in this process used for?

  ___to make the body move and make substances in the body.___

10 What is oxidation? Tick (✔) the correct box.

A process in which oxygen is taken away from a substance. ☐

A process in which oxygen is added to a substance. ☒

The bubbling of oxygen through a liquid. ☐

The breakdown of ozone in the atmosphere. ☐

11 Here are the stages in the rusting of iron but they are in the wrong order. Arrange them correctly by writing the letter of each statement in the order in which it occurs.

A Oxygen reacts with iron to form iron oxide.

B Water vapour condenses on iron.

C Brown flakes fall away and the metal corrodes.

D Oxygen dissolves in the layer of condensed water.

E Iron oxide forms brown flakes.

B D A E C

12 Draw lines to match each part of a hand warmer to its functions.

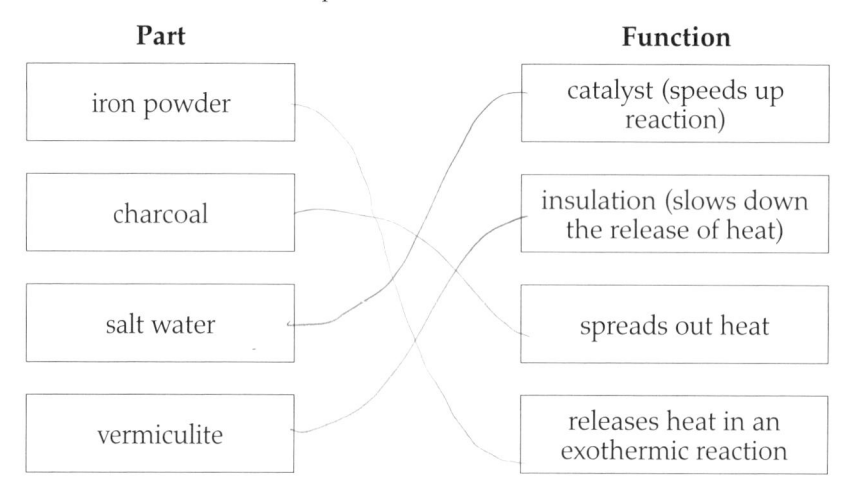

Part	Function
iron powder	catalyst (speeds up reaction)
charcoal	insulation (slows down the release of heat)
salt water	spreads out heat
vermiculite	releases heat in an exothermic reaction

Teacher comments

10 Patterns of reactivity

1. a) How is sodium stored? _in oil_

   b) Explain your answer. _it has to be kept there because it is very reactive and will rapidly react with oxygen and water_

Reaction of metals with oxygen

2. Use the following words and phrases in your answers to the parts of this question.
   - black powder
   - glows
   - not changed
   - makes yellow sparks
   - forms black powder on the surface

   What happens when each of the following metals is heated?

   a) Iron _glows, makes yellow sparks and forms black powder_

   b) Gold _is not changed_

   c) Copper _forms black powder on the surface_

● CHAPTER 10

Reaction of metals with water

3

Diagram labels:
- rocksil wool soaked in water
- HEAT
- metal sample (magnesium, iron, zinc)
- rubber bung
- hydrogen
- water

a) What is this apparatus used for? __To investigate the reaction between a metal and water steam__

b) Label the parts of the diagram.

c) Draw an arrow and write the word 'heat' where you think it is applied to the apparatus.

d) (i) Which gas is collected in this reaction? __hydrogen__

   (ii) What is the test for this gas? __if there is hydrogen it combines with oxygen in the air and makes an explosive popping sound.__

e) Which of the following metals would produce a reaction if placed in this apparatus? Circle your answers.
   - gold
   - **magnesium**
   - **zinc**
   - silver
   - **aluminium**
   - **calcium**
   - copper
   - tin

60

Reaction of metals with acids

4 Jaafar is selecting apparatus to investigate the reaction of some metals with an acid.

   a) Which of the following pieces of apparatus should he choose? Circle your answers.
   - trough
   - thermometer
   - Bunsen burner
   - delivery tube
   - thistle funnel
   - Buchner funnel
   - bung with two holes
   - bung with one hole
   - beaker
   - test tube
   - tripod
   - gauze
   - conical flask

   After he has assembled the apparatus, put in the metal and added the acid there is a fizzing and a gas is produced.

   b) (i) What is this gas? _____

   (ii) Where is it collected? _____

   Jaafar has used hydrochloric acid in his investigation and has begun to write the word equation.

   c) Complete the word equation.

   metal + hydrochloric acid → _____

   d) Jaafar decides to set up the apparatus again and try a second metal. How could he find out whether the metal was more reactive or less reactive than the one he had just investigated?

   _____

   _____

   _____

   e) If he tested the following metals, which ones would he see react with the acid? Circle your answers.
   - silver
   - copper
   - magnesium
   - aluminium
   - calcium
   - gold
   - zinc

CHAPTER 10

5 When a copper wire is placed in a solution of silver sulfate, the solution turns blue and silver appears on the wire.

   a) Why does the solution turn blue? _____

   b) Where does the silver come from? _____

   c) Why does this change take place? _____

   d) (i) If an iron nail was placed in the copper sulfate solution, would you see a change? _____

   (ii) Explain your answer. _____

6 Here are some metals that are in the reactivity series. List the top **five** in the correct order.
   - copper
   - potassium
   - calcium
   - tin
   - zinc
   - iron
   - gold
   - aluminium
   - sodium
   - magnesium

   1. Potassium
   2. Sodium
   3. Calcium
   4. Magnesium
   5. Aluminium

Teacher comments

11 Preparing common salts

1. Identify **two** uses of calcium chloride. Tick (✔) the two correct boxes.

   In herbicides ☐
   As a drying agent ✔
   Setting concrete ✔
   Bleaching paper ☐

2. Identify **two** uses of zinc sulfate. Tick (✔) the two correct boxes.

   Making cosmetics ✔
   Food processing ☐
   In sewage treatment ✔
   Setting concrete ☐

Acids and their salts

3. Draw lines to match each salt with its colour.

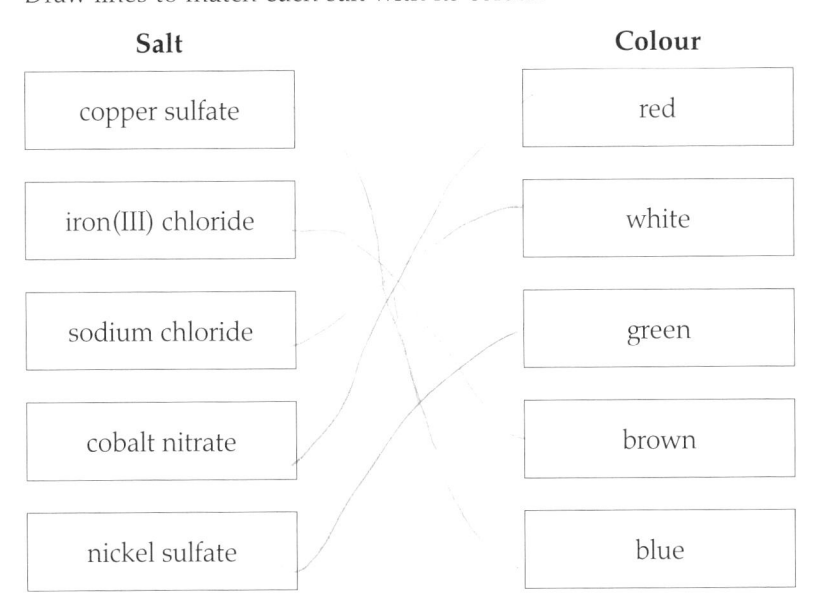

Salt	Colour
copper sulfate	red
iron(III) chloride	white
sodium chloride	green
cobalt nitrate	brown
nickel sulfate	blue

● CHAPTER 11

4 The table lists the apparatus used in the preparation of salts by reacting metals and metal carbonates with an acid.

Apparatus number	Apparatus name	Use of apparatus
1	Bunsen burner	A
2	conical flask	H
3	filter paper	F
4	clamp and stand	I
5	gauze	C
6	evaporating dish	J
7	spatula	D
8	tripod	E
9	beaker	B
10	filter funnel	G

a) Match the use of the apparatus below with their names by writing the letters in the appropriate box in the table.

Letter	Use
A	provides heat
B	collects filtrate
C	supports evaporating dish
D	transfers metal to acid
E	supports gauze
F	separates unreacted metal from a liquid
G	holds filter paper
H	container where acid and solid are mixed
I	holds filter funnel
J	where solid salt forms

64

PREPARING COMMON SALTS

b) Here are the stages in the preparation of a salt from a metal or metal carbonate but they are in the wrong order. Arrange them correctly by writing the letter of each statement in the order in which it occurs.

    A Look for the end of bubbling in mixture.

    B Add solid to acid.

    C Leave heated solid to cool.

    D Heat liquid in heatproof container until solid appears.

    E Pour acid into container.

    F Transfer separated liquid to a heatproof container.

    G Separate solid from liquid.

    E B A G F D C

c) At the end of this sequence, what further process is performed to remove any remaining liquid? *the substance is filtered again*

5 Han is preparing copper sulfate using sulfuric acid. He has a sample of copper metal and a sample of copper carbonate.

    a) Which sample should he use? *copper metal*

    b) Explain your answer. *so that no carbon dioxide is released*

Teacher comments

12 Rates of reaction

1. What does **rate** mean? Tick (✔) the correct box.

   A measure of the rise of temperature in a reaction ☐

   A measure of the amount of chemicals in a reaction ☐

   A measure of the speed of change in a reaction ☑

   A measure of the concentration of the reactants ☐

Measuring rate of reaction

2. Sarah is planning to measure the change in mass during a reaction.

   a) What piece of apparatus does she need to do this? _A balance_

   b) Here are some instructions for her to follow but they are in the wrong order. Arrange them correctly by writing the letter of each statement in the order in which it occurs.

   A Find the mass of mixed reactants.

   B Plot the data collected on a graph.

   C Find the mass of the separate reactants.

   D Record the mass of the reactants frequently over a period of time.

   E Put the reactants together.

   _C A E D B_

   c) After her first investigation, Sarah decides she will measure the volume of a gas produced in a reaction. What special piece of apparatus will she need for this investigation? _a gas syringe_

Factors affecting rates of reaction

3 What is the concentration of a liquid? Tick (✔) the correct box.

A measure of the amount of solvent in the solution ☐

A measure of the amount of solute in the solution ✔

A measure of the amount of solvent in the solute ☐

A measure of the amount of solution in the solvent ☐

4 Shazia takes a lump of reactant and adds it to another substance and records the time when the reaction stops. It takes 5 minutes. She then grinds up another lump of reactant, the same size as the first, with a pestle and mortar.

   a) What is happening to the size of the particles she makes as she keeps grinding the reactant? _they are getting smaller_

   b) What is happening to the surface area of the reactant as she grinds it down?

   It is increasing.

   c) (i) If Shazia repeats the experiment with the powder, how long will the reaction take? Tick (✔) the correct box.

   More than 5 minutes ☐

   5 minutes ☐

   Less than 5 minutes ✔

   (ii) Explain your answer. _There is more surface area of the reactant in contact with the substance so it will react quicker_

5 David is investigating the volume of gas produced in a reaction. Here are his results.

Time/minutes	Volume/cm^3
0	0
1	6
2	12
3	18
4	22
5	24
6	25
7	25

CHAPTER 12

a) Plot the data on the graph.

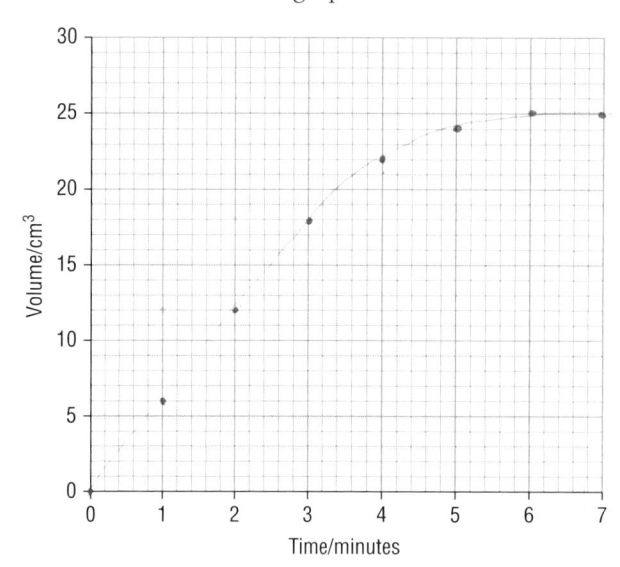

[Graph: Volume/cm³ vs Time/minutes, with plotted points showing a curve rising from 0 to approximately 25 cm³ and levelling off]

The laboratory temperature was 25 °C. The teacher switches on the air conditioning to make it cooler. David repeats the investigation and plots a second graph.

b) (i) Where will the second graph be? Tick (✔) the correct box.

At the same place as the first graph ☐

Above the line of the first graph ☐

Below the line of the first graph ✔

(ii) Explain your answer. <u>The reaction will take longer so the gradient will be less steep.</u>

Catalysts

6 Here are some statements that could describe catalysts. Identify the ones you think are correct by putting a tick (✔) in appropriate boxes.

Statement	Tick if correct
A catalyst only speeds up one reaction.	✓
A catalyst speeds up lots of reactions.	
A small amount of catalyst is needed to speed up a reaction.	✓
A large amount of catalyst is needed to speed up a reaction.	
The catalyst is chemically changed in the reaction.	
The catalyst is not chemically changed in the reaction.	✓
A catalyst cannot be used again after taking part in a reaction.	
A catalyst can be used again after taking part in a reaction.	✓

7 Hydrogen peroxide breaks down naturally into two substances and one of them is a gas.

   a) Write the word equation for this reaction.

   hydrogen peroxide → water + oxygen

   b) What is the catalyst that speeds up this reaction? manganese oxide

   c) If the gas is collected, what test would you perform on it to identify it and what would you expect to see? putting a splint in the test tube where the gas is. I would expect the flame to keep burning.

8 a) Name three gases produced by car engines. carbon monoxide, hydrocarbons, nitrous oxides.

   b) State one way in which one of the gases is harmful. when carbon monoxide is inhaled, it stays in your red blood cells.

   c) Name the elements used to make the catalyst in a catalytic converter.

   platinum and rhodium

   d) Name three substances released by a car after its exhaust gases have passed through a catalytic converter. water, carbon dioxide, nitrogen

● CHAPTER 12

9 a) What is the name of catalysts in living things? __enzymes__

   b) From what are these catalysts made? __proteins__

   c) Name a place in the body where they are found. __liver__

   d) What do these catalysts in biological washing powders break down?

   __fat and protein stains.__

The particle theory and rates of reaction

10 a) For a reaction to take place, what must the particles do?

   __collide with each other__

   b) (i) If a solution is diluted, what happens to the number of particles that take part in a reaction? __it decreases because it is less concentrated__

   (ii) How does this affect the rate of reaction? __it decreases it__

   c) (i) What effect does raising the temperature have on the movement of particles in a reaction? __the particles move faster__

   (ii) How does this change the rate of reaction? __it increases it__

   (iii) Why does it change the rate of reaction? __because moving faster makes harder collisions which are more likely to react__

   d) How does a catalyst affect the particles taking part in a reaction?

   __They provide a surface for them to settle and react, then move away for other particles to react.__

Teacher comments

PHYSICS

13 Density

Comparing densities

1 Dembe has three small logs of different wood.

   a) What must he do to compare their densities? find the masses of the logs with equal volumes

   b) Which formula should he use to calculate the density of each wood? Tick (✔) the correct box.

   density = mass × volume ☐

   density = volume/mass ☐

   density = mass/volume ✔

2 Jaya has a block of material and wants to find its volume.

   a) How should she do this? divide its mass by its density (also: length × height × width)

   b) Jaya's block is a cube with sides of 5 cm. What is its volume? Show your working.

   $5 \times 5 \times 5 = 5^3 = 125$ $125 \, cm^3$

● CHAPTER 13

3 a) Shazad wants to find the density of a pebble. Here are the stages he should use but they are in the wrong order. Arrange them correctly by writing the letter of each statement in the order in which it occurs. The first statement has been placed for you.

   A Put pebble on a balance.
   B Attach a string to the pebble.
   C Half fill a measuring cylinder with water.
   D Read the volume of water and pebble. (V2)
   E Carefully lower the pebble into the measuring cylinder.
   F Make sure that the pebble is completely covered by water.
   G Read off mass of pebble. (M)
   H Read the first level of water. (V1)

   A _G C H B E F D_

   b) How should Shazad find the volume of the pebble? _by subtracting the V2 of the V1._

   c) Shazad's pebble has a mass of 90 g and a volume of 30 cm³. What is its density? Show your working.

   $\rho = \dfrac{90g}{30 cm^3}$ $\rho = 3 g/cm^3$

   $\rho = \dfrac{M}{V}$

4 Fozia wants to find the density of honey.
   a) Here are the stages she should use but they are in the wrong order. Arrange them correctly by writing the letter of each statement in the order in which it occurs.

   A Read the mass of the measuring cylinder. (M1)
   B Read the mass of the cylinder and liquid. (M2)
   C Read the volume of liquid in the measuring cylinder. (V)
   D Put the measuring cylinder on the balance.
   E Place the cylinder containing the liquid on the balance.
   F Pour the liquid into the measuring cylinder.

   _D A F B E C_

b) Which formula should Fozia use to calculate the density of the honey? Tick (✔) the correct box.

Density = M1 – M2/V ☐

Density = M2 – M1/V ✔

Density = M1 + M2/V ☐

Density = V – M1/M2 ☐

c) Fozia's results are A 100 g, B 120 g and V 14 cm³. What is the density of the honey? Show your working.

$$\rho = \frac{m_2 - m_1}{V} \quad \rho = \frac{120g - 100g}{14 cm^3} \quad \rho = \frac{20g}{14 cm^3} \quad \rho \approx 1.43 g/cm^3$$

Floating and sinking

5 Budi has three liquids: water, vegetable oil and maple syrup. He says the water has a density of 1, the vegetable oil has a density of 0.92 and the maple syrup has a density of 1.37.

a) What are the units he is using to measure the density?

g/cm^3

b) What would the units be if he multiplied them by a thousand?

kg/m^3

Budi pours each one into the same tall jar. They settle out in layers.

c) (i) Which liquid is at the top? __oil__

(ii) Which liquid is in the middle? __water__

(iii) Which liquid is at the bottom? __maple syrup__

(iv) Explain your answer. __oil is the least dense so it will float to the top, maple syrup is the most dense so it will sink to the bottom__

● CHAPTER 13

d) Budi pours some corn oil into the jar. He says it has a density of 0.97. Where does it settle? Tick (✔) the two correct boxes.

Above the water ☑

Below the vegetable oil ☑

Above the maple syrup ☐

Below the maple syrup ☐

Density of gases

6 A teacher is demonstrating how to find the density of air.

a) Here are the stages she uses but they are in the wrong order. Arrange them correctly by writing the letter of each statement in the order in which it occurs.

A Put the evacuated flask on the balance.

B Pour water from the flask into a measuring cylinder.

C Find the mass of the flask. (M1)

D Open the flask underwater.

E Measure the volume of the water. (V)

F Place the flask on the balance.

G Let the water replace the vacuum in the flask.

H Measure the mass of the evacuated flask. (M2)

I Remove air from the flask with a vacuum pump.

F C I A H D G B E ✓

b) Which formula should the teacher use to calculate the density of the gas? Tick (✔) the correct box.

Density = M1 – M2/V ☑ ✓

Density = M2 – M1/V ☐

Density = M2 – V/M1 ☐

Density = M1 – V/M2 ☐

DENSITY

7 a) The particles of a gas move further apart when they are heated. How does this affect the density of the gas? _the density decreases_ ✓

b) The particles of a gas are pushed closer together when the pressure on the gas is increased. How does this affect the density? _the density ~~also~~_ ✓ _increases because the mass increase in comparison to the volume._

c) If the densities of two gases are measured at the standard temperature and pressure (STP), what does it mean? _that both gasses have ~~temperature~~ ~~are~~ are measured at the same temperature (0°C) and pressure (will support 760 mm of mercury)_

Teacher comments

⚠ gasses' densities need to be measured in the same temperature and pressure.

14 Pressure

1. Name **four** things that a force can do. change direction, change speed, change side, start object moving, stop object moving. ✓

Pressure on a surface

2. What is weight? Tick (✔) the correct box.

   A force pushing out in all directions ☐

   The amount of matter in an object ☐

   The force produced by gravity acting on an object ☒ ✓

   The volume of matter in an object ☐

3. A block has three surfaces labelled A, B and C.

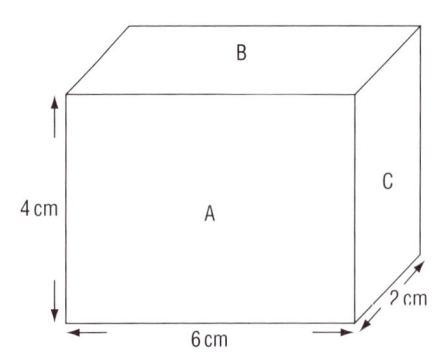

   a) What is the area of each of the three surfaces? Show your working.

   A 4 × 6 = 24   24 cm² ✓

   B 6 × 2 = 12   12 cm² ✓

   C 2 × 4 = 8   8 cm² ✓

   The weight of the block is 48 N.

PRESSURE

b) What is the pressure on the ground when the block is stood on each surface? Show your working.

A $48/24 = 2$ 2 N/cm^2 ✓

B $48/12 = 4$ 4 N/cm^2 ✓

C $48/8 = 6$ 6 N/cm^2 ✓

4 Two identical cows fell into a muddy hole. One fell on its side and the other fell on its feet.

a) Which one sank furthest into the mud? _the one on its feet_ ✓

b) Explain your answer. _the pressure is higher because there is a lower surface area._ ✓

5 Ingrid goes outside to see if the snow is fit for skiing. She sinks into the snow but when she puts her skis on, she can move over it without sinking. Why?

because the pressure is distributed on more surface area, so the pressure pushing on the snow is lower. ✓ (and weight doesn't change) !

6 Paulo is playing football in his trainers but keeps slipping. He changes to boots with studs and stops slipping. Why? _because the boots are putting more pressure and touching less area on the flour_ ✓

7 A drawing pin has a head and a point. When you push it into a board which part is under:

a) high pressure _point_ ✓

b) low pressure? _head_ ✓

8 A chef is having difficulty cutting up onions because his knife is blunt. After he sharpens it, the knife cuts more easily. Why is this?

the knife now has a lower surface area on its edge ✓

● CHAPTER 14

Pressure in liquids

9 A beaker contains a liquid. State **two** places where the liquid exerts a pressure.

   bottom and sides ✓

10 The picture shows three jets of water flowing from a can.

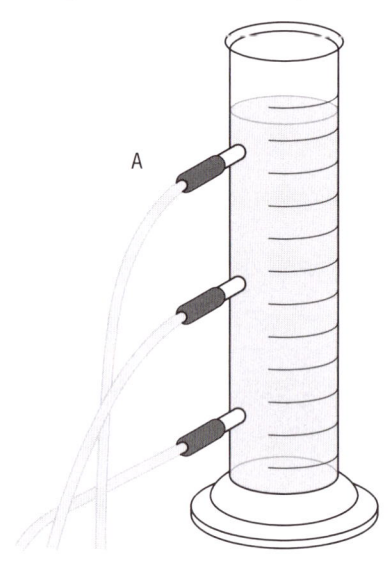

   a) When will jet A stop flowing? _when there's no longer water above it._ ✓
   b) As the water level falls, what happens to the middle jet of water?
      Tick (✔) the correct box.

      It stays the same.   ☐

      It rises and becomes more horizontal.   ☐

      It sinks and becomes more vertical.   ☑ ✓

11 a) What is hydraulic equipment used for? Tick (✔) the correct box.

      Transmitting liquids from one place to another   ☐

      Transmitting pressure from one place to another   ☑ ✓

      Transmitting pressure in all directions   ☐

      Transmitting pressure from one liquid to another   ☐

      (small force → big force)

b) Name **two** uses of hydraulic systems. _brake system of a car,_
Car raising thing for repair.

Pressure in gases

12 A teacher is demonstrating the strength of the air pressure by using a steam can. Here are the events that occur in the demonstration but they are arranged in the wrong order. Arrange them correctly by writing the letter of each statement in the order in which it occurs.

  A The top is put on the can.
  B The can is crushed.
  C A small amount of water is poured into a can.
  D Steam comes out of the can.
  E The quantity and pressure of the air inside the can is reduced.
  F The can is heated.
  G Condensation of steam occurs inside the can.
  H The heat is switched off.

  _C F D H A G E B_ ✓

13 **a)** Why does the gas shoot out of an aerosol when its nozzle is opened?

  there is a higher pressure than in the air, so it pushes ✓
  the liquid out.

An aerosol can has strong sides that keep their shape when the can is full.

  b) (i) What would happen if the sides were weak?

  _____

  (ii) Explain your answer. _____

  _____

● CHAPTER 14

14 a) What is used to collect the air under a hovercraft?

powerful fans

b) What prevents the air escaping? _the skirt at the edge_

c) What makes the hovercraft rise? _upward pressure made_

Teacher comments

15 Turning on a pivot

1. Which statement best describes a lever? Tick (✔) the correct box.

   It changes the size of a force. ☐

   It changes the direction of a force. ☑

   It changes the size or direction of a force. ☐

   It changes the area affected by a force. ☐

The turning effect of forces

2. Shazia says that the fulcrum is part of a lever and Ali says the pivot is part of a lever.

   a) (i) Who is correct? _both_

   (ii) Explain your answer. _they mean the same think_

   Ali asks Shazia what she thinks fulcrum means.

   b) What should she reply? _the fulcrum is the object ✓ under supporting the laver the arm or two arms_

   Shazia says that there are also two other parts to a lever.

   c) What are they? _two arms_

3. What is the difference between the load and the effort?

   The load is _the force resisting the moment_

   The effort is _the force applied_

● CHAPTER 15

Types of levers

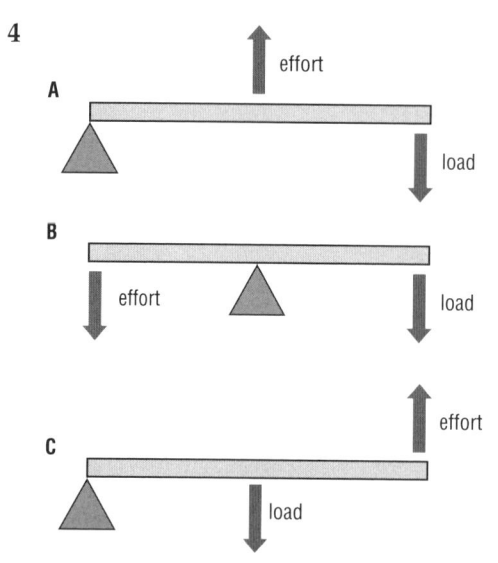

4

A — effort (up, middle), load (down, right), fulcrum (left)

B — effort (down, left), fulcrum (middle), load (down, right)

C — fulcrum (left), load (down, middle), effort (up, right)

Name the class of each lever and give an example.

Lever	Class	Example
A	third	broom
B	first	seesaw
C	second	nut cracker

Moments

5 What is a moment? Tick (✔) the correct box. A moment is a measure of the turning effect produced by a force:

around a point supporting the lever. ✔

around the ends of the lever. ☐

around the effort of the lever. ☐

around the load of the lever. ☐

6 Who constructed the law of moments? Tick (✔) the correct box.

Aristotle ☐

Democritus ☐

Galileo ☐

Archimedes ☐

7 Here are some parts of the definition of the law of moments but they are in the wrong order. Arrange them in the correct order by writing them out in the space provided.
- the sum of the clockwise moments about any point
- is in equilibrium
- the anticlockwise moments about that point
- equals the sum of
- when a body
- (or balanced)

When a body is in equilibrium (or balanced), the sum of the clockwise moments about any point equals the sum of the anticlockwise moments about that point.

8 Shen found a long plank and set it up as a seesaw. His weight is 450 N and he sits 2 metres from the point on which it turns. How far will his younger sister Bo, who weighs 300 N, have to sit from the point to balance her brother? Show your working.

moment = force × distance   450 N × 2 m = 900 Nm

300 N × 3 m = 900 Nm   [3 m away]

Teacher comments

16 Electrostatics

The atom and electric charge

1. Which parts of the atoms of a material move to generate a charge on it? Tick (✔) the correct box.

    Nuclei ☐

    Protons ☐

    Neutrons ☐

    Electrons ☐

Charging materials

2. A balloon is rubbed on a woollen sleeve.

    a) What happens to the sleeve? _____

    b) What happens to the balloon? _____

3. Two charged balloons are hung from threads and are brought close together.

    a) What happens to the balloons? _____

    b) Explain the observation. _____
    _____

4. A piece of Perspex is rubbed on a woollen sleeve.

    a) What happens to the sleeve? _____

    b) What happens to the Perspex? _____

ELECTROSTATICS

5 A charged piece of Perspex is brought near a charged balloon hanging on a thread.

   a) What happens to the balloon? _____

   b) Explain the observation. _____

Insulators and conductors

6 How are insulators and conductors different? _____

Induced charges

7 A charged balloon induces a charge on the surface of a wall.

   a) What is the charge on the balloon? _____
   b) What happens at the surface of the wall when the charged balloon induces a charge?

   _____

   c) What is the induced charge on the wall surface? _____
   When the balloon touches the wall and is released, it sticks to the wall.

   d) What force holds the balloon to the wall? _____

   e) (i) How does this force compare with the force of gravity between the balloon and the centre of the Earth? _____

      (ii) Explain your answer. _____

CHAPTER 16

Sparks and flashes

8 Here are the events that happen when a spark occurs but they are in the wrong order. Arrange them correctly by writing the letter of each statement in the order in which it occurs.

   A Electrons split more molecules as they move through the air.
   B The size of two oppositely charged surfaces becomes very large.
   C Electrons and ions meet charged surfaces and the charges cancel out.
   D Electrons and ions are formed.
   E Electrons and ions move to the charged surfaces.
   F The molecules in the air are split.
   G The huge number of electrons and ions make a spark.

9 a) How does an aircraft become charged as it flies? _____

   b) How is the aircraft discharged on landing? _____

10 What is the difference between sheet lighting and forked lighting?

   Sheet lightning _____

   Forked lightning _____

11 What is the name of the machine that is used to generate huge electrostatic charges? _____

12 Digital sensors and touch screens could not have been developed without the invention of a charge-storing device. What is it called? _____

Teacher comments

17 Electricity

Simple circuits

1 Naveen sets up the circuit shown in the diagram.

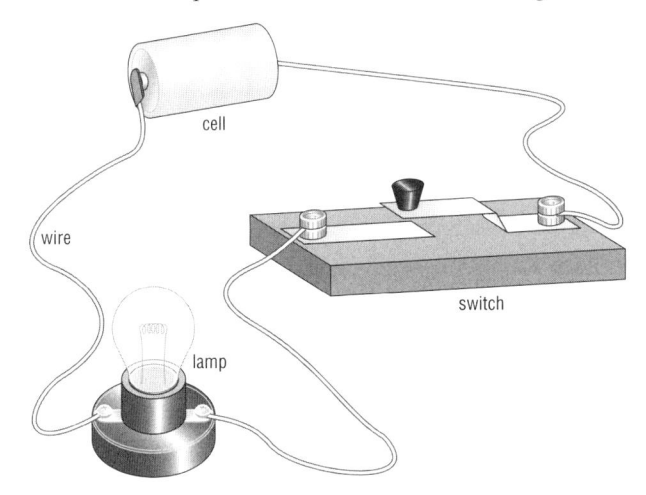

   a) When he closes the switch, which of the following happens in the wire? Tick (✔) the correct box.

   The electrons move from the negative terminal to the positive terminal. ☐

   The atoms move from the negative terminal to the positive terminal. ☐

   The electrons move from the positive terminal to the negative terminal. ☐

   The atoms move from the positive terminal to the negative terminal. ☐

   b) Where does the energy to create the current come from? Circle your answer.
   - wire
   - cell
   - lamp
   - switch

   c) As the current flows through the circuit, the wire in the lamp behaves differently from the other wires in the circuit. What happens to the wire?

   _____

   Naveen leaves the circuit on for some time and later observes that the current has become weaker. →

d) How can he tell just by looking at the circuit? _____

e) What has caused this change in the current? _____

_____

Naveen opens the switch.

f) (i) What happens to the current? _____

(ii) Explain your answer. _____

_____

2 Here are the symbols for the components in Naveen's circuit. The one on the left is for the cell.

a) Put a cross (X) next to the side which has the negative terminal.

b) Make a circuit diagram of Naveen's circuit using these symbols.

Naveen adds another cell in series to the circuit.

c) Draw the circuit diagram for this new circuit.

Naveen switches on the circuit.

d) (i) How does the appearance of the lamp change? _____

(ii) Explain your answer. _____

_____

3 The connection of the wire to the lamp breaks. Aruni suggests putting a piece of aluminium foil across the gap. Naveen disagrees and says a piece of wood is all that is needed.

a) Who is correct? _____

b) Explain your answer. _____

_____

CHAPTER 17

Resistance and Other circuit components

4 Here are the symbols of three more components of circuits. What are their names?

    A   B   C

A _____ B _____

C _____

5 a) What happens when a fuse wire gets very hot? _____

    b) How does this change affect the flow of current? _____

    c) What causes a fuse wire to get very hot? _____

    d) Why is a fuse a safety device? _____

    e) How is a circuit breaker different from a fuse? _____

Amperes

6 a) What does an ammeter measure? _____

    b) An ammeter has its positive terminal marked in red. When adding an ammeter to a circuit, which instruction should you follow? Tick (✔) the correct box.

      Connect red terminal to positive terminal of cell ☐

      Connect red terminal to negative terminal of cell ☐

      Connect red terminal to either terminal of cell ☐

      Connect red terminal to a lamp ☐

7 Ochi sets up the circuit shown in the diagram. She connects the ammeter at point A and then at point B.

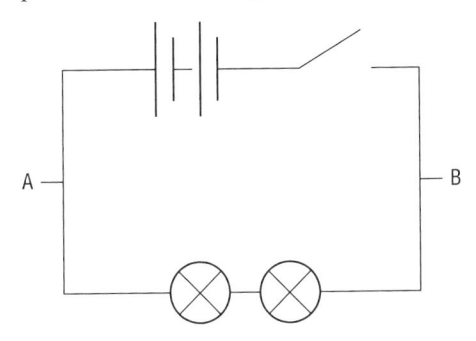

What does she find? Tick (✔) the correct box.

The reading at A is lower than at B. ☐

The reading at A is higher than at B. ☐

The reading at A is the same as at B. ☐

8 Kali sets up the circuit shown in the diagram.

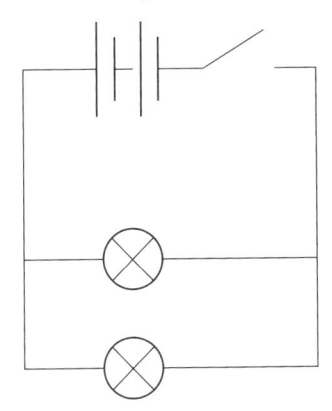

a) How many places in the circuit should she check with the ammeter? Circle your answer.
- 2
- 3
- 4
- 5

b) Mark the places on the circuit diagram.

● CHAPTER 17

Voltage

9 What does the voltage measure? Tick (✔) the correct box.

A difference in potential energy ☐

A difference in current speed ☐

A difference in resistance ☐

A difference in lamp brightness ☐

10 The diagram shows the symbol for the voltmeter and a simple circuit.

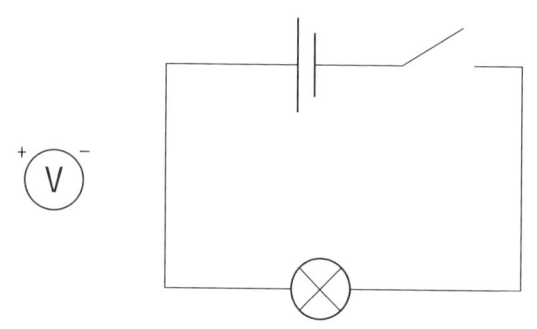

Draw how you would connect the voltmeter to the circuit to measure the voltage across the lamp.

Teacher comments

18 Heat energy transfers

Heat and internal energy

1. What is heat a measure of? Tick (✔) the correct box.

   The mass of atoms and molecules in a substance. ☐

   The kinetic energy of the atoms and molecules in a substance. ☐

   The potential energy of the atoms and molecules in a substance. ☐

   The different kinds of atoms and molecules in a substance. ☐

2. How do the particles move in a hot substance? Tick (✔) the correct box.

   Slower than in a cold substance ☐

   At the same speed as those in a cold substance ☐

   Faster than in a cold substance ☐

   In a different direction to those in a cold substance ☐

3. A scientist sets out to compare how three materials conduct heat. She cuts each material into a strip. She puts one end by a heat source and connects a temperature sensor of a data logger to the other end.

   a) How should she make the test fair? _____

   _____

   _____

   _____

CHAPTER 18

Here are the results for the change in temperature recorded by the data logger.

Time/seconds	Material A Temperature/°C	Material B Temperature/°C	Material C Temperature/°C
0	20	20	20
15	22	27	24
30	24	34	30
45	26	41	32
60	28	48	34
75	30	55	40
90			

b) Plot the data for materials A and B on the graph.

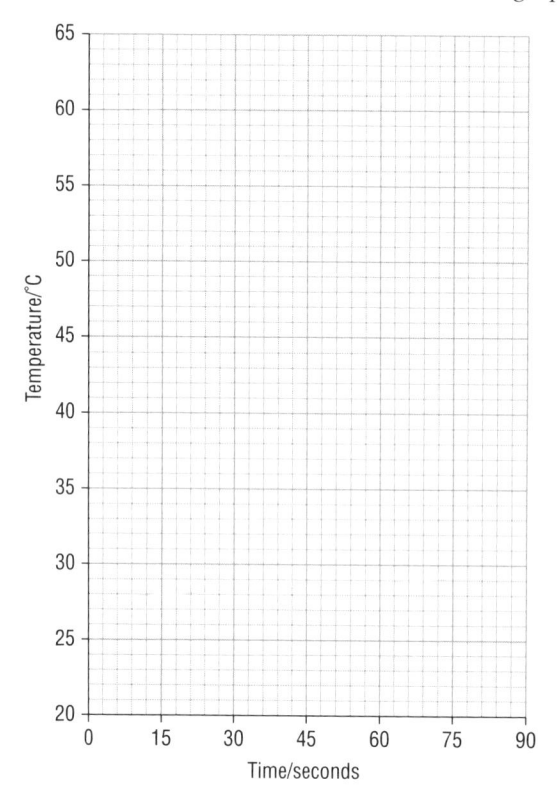

c) Predict the temperature of each one at 90 seconds and record them in the table.

d) Plot the data for material C but do not join up the points. Look for the line of best fit and draw it in.

HEAT ENERGY TRANSFERS

e) Use this line to predict the temperature of C at 90 seconds and record it in the table.

f) Suggest reasons for the results in the test with material C. _____

_____

_____

4 Ingrid and Karl are investigating the heat-conducting properties of two materials.
   a) Which apparatus will they need for their investigation? Circle your answers.
   - beaker
   - newton meter
   - tripod
   - thermometer
   - kettle
   - Bunsen burner
   - gauze
   - measuring cylinder
   - test tube
   - delivery tube
   - Liebig condenser
   b) How should they set up each material for the investigation?

   _____

After the investigation Karl takes each material and puts it under water and squeezes it. Material B releases more bubbles than material A.

c) (i) Which material do you predict will have kept most heat in the apparatus?

   _____

   (ii) Explain your answer. _____

   _____

● CHAPTER 18

5   Lily lights a joss stick and watches the smoke. It rises in the air, then spreads out and sinks. Explain this observation. _____

_____

_____

_____

6   a) How is energy carried by radiation? Tick (✔) the correct box.

   Sound waves   ☐

   Electrons   ☐

   Electromagnetic waves ☐

   Electrostatic charges ☐

   b) Why can heat pass through a vacuum by radiation but not by conduction or convection? _____

   _____

7   a) Chan has painted one tin can black and left another one shiny. He has a thermometer, measuring cylinder and a bottle of water. How could he find out which surface absorbs most heat from the Sun?

   _____

   _____

   _____

   _____

   _____

HEAT ENERGY TRANSFERS

b) (i) What do you predict will be the result? _____

(ii) Explain your answer. _____

_____

Evaporation

8   Abdul's face is covered in sweat after running.

a) What happens at the surface of the water in his sweat? _____

_____

b) What happens to the mass of water in the sweat? _____
c) What happens to the amount of energy in the water in the sweat?

_____

d) What happens to the temperature of the water in the sweat?

_____

9   In the cool compartment of a fridge there are pipes carrying liquid from a compressor. The pressure in the pipes is much lower than in the compressor.

a) How does the fall in pressure affect the liquid? _____

b) What happens to the number of particles in the liquid? _____

The change in the liquid makes the pipes cooler than the air in the compartment. Heat is drawn from the air through the wall of the pipe.

c) What form of thermal energy transfer takes place? _____

Teacher comments

19 World energy needs

Energy pathways on Earth

1 These statements describe the fate of most of the energy from the Sun as it reaches the Earth but they are in the wrong order. Arrange them in the best order by writing the letter of each statement in the order in which it occurs.

   A Some energy is used in photosynthesis in leaves.
   B Some energy produces winds and water currents.
   C Some energy travels from the Sun through space.
   D Some energy is absorbed by the atmosphere.
   E Some energy is absorbed by the surface of the Earth.
   F Some energy is reflected back into space by the atmosphere.
   G Some energy is used to power the water cycle.

The water cycle

2 The diagram shows some parts of the water cycle. Fill in the other parts and draw in arrows to show the path of water through the cycle.

Winds, waves, tidal and geothermal energy

3   Draw lines to match the energy sources to their origins.

Energy source	Origin
tidal	radioactive materials
geothermal	convection currents
wind	moving air on a surface
waves	gravity

Fossil fuels

4   a) Where does the energy in fossil fuels originally come from? _____

   b) Coal is a fossil fuel. What kind of coal mining causes the most habitat destruction? _____

   c) Name **two** more fossil fuels. _____

   The extraction and transport of one fossil fuel can damage aquatic habitats.

   d) (i) Name the fossil fuel. _____

      (ii) Name the type of aquatic habitats that are damaged. _____

Non-renewable fuels

5   Fossil fuels, such as coal, are non-renewable energy sources.

   a) What does this mean? _____

   b) Name another non-renewable energy source. _____

● CHAPTER 19

Renewable energy sources

6 What does the term **renewable energy source** mean? _____

_____

7 Two sorts of devices have been made for trapping solar energy. Name the **two** devices and the form of energy they trap.

Device   **Form of energy**

_____   _____

_____   _____

8 a) What kind of motion in waves can be used to generate useful energy?

_____

   b) In what form is the useful energy? _____

9 a) What form of energy is used as the energy source in rivers and the wind? _____

   b) What is the name of the devices used to convert this form of energy into a more useful form? _____

   c) What form of energy do these devices produce? _____

Teacher comments